*"Jordan's teaching of the facts
tactful an*
— Nikki Gonzalez, Executive Di

THE
PROCEDURES

A Quick Read Regarding Abortion Procedures

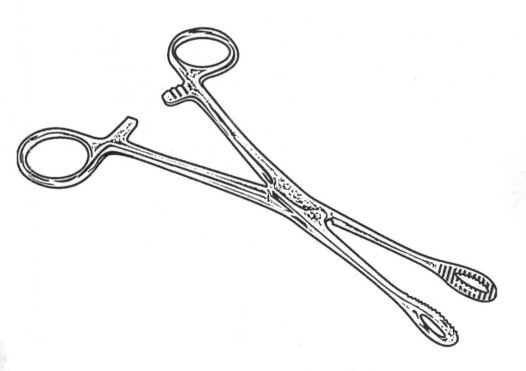

JORDAN LYNN WARFEL

The Procedures

Copyright © 2020 by Jordan Lynn Warfel.

All rights reserved.
No part of this book may be reproduced in any form or by any electronic or mechanical means, including information storage and retrieval systems, without permission in writing from the author. The only exception is by a reviewer, who may quote short excerpts in a review.

Illustrations by Marco Chiu
Cover design by PixelStudio
Layout design by Inkcept Studio
Editing by Natalie Groff

This book includes detailed descriptions of abortion procedures. Reader discretion is advised. This book does not contain pictures of aborted children. Nothing in this book should be construed as medical or legal advice. If you are struggling with the loss of a child, please seek professional help.

Publisher is not responsible for websites or their content that are not owned by the publisher.

First paperback edition: April 2020
ISBN: 978-1-7344108-2-2
Candlelight Publishing
Greenwood, Delaware

Jordan Lynn Warfel

THE PROCEDURES

A Quick Read Regarding Abortion Procedures

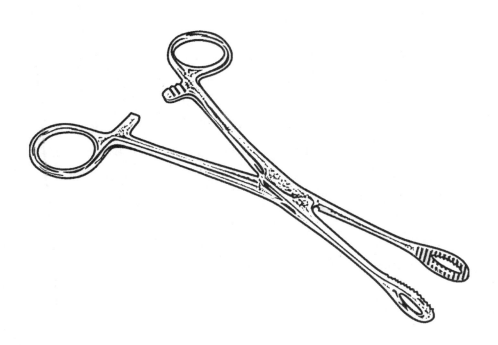

CANDLELIGHT PUBLISHING
GREENWOOD

ABOUT THE AUTHOR

Jordan Warfel resides in Delaware with his wife Felicia and their three children. Jordan was born and raised in southern Delaware where his parents were board members for the local pro-life pregnancy care center. After graduation, he achieved an associate of arts degree in Biblical studies from Rosedale Bible College and a bachelor of science degree in organizational management from Wilmington University. Since 2005 Jordan has been involved with numerous aspects of the pro-life movement including legislation, political campaigns, and media. Today Jordan is the president of the annual Candlelight Walk for Life in Milford, Delaware and regularly gives pro-life presentations to churches and community groups. When he is not working as a residential draftsman and in the pro-life movement, he enjoys playing with his kids and making improvements to their home.

SPEAKING REQUESTS

Jordan Warfel is available to speak to your church, school, or community group. You can request a hands-on interactive presentation for audiences of all sizes based on the content in *The Procedures*. Presentations are affordable and can be tailored to the interests of your audience. To book a presentation, send an email to jordanwarfel@gmail.com.

www.Candlelight.LIFE

DEDICATION

This book is dedicated to Nicole,
for entrusting us to adopt your son.
Our society is brimming with messages telling
you that abortion is the solution.
But you had the courage to do
what was best for your unborn child.
We owe a debt to you that we can never repay.
When we think of heroes,
we think of you.

CONTENTS

Chapter 1	The Curette	1
Chapter 2	The Vacuum	19
Chapter 3	The Abortion Pill RU486	33
Chapter 4	Abortion Pill Reversal	49
Chapter 5	The Forceps	61
Chapter 6	The Lethal Injection	75
Chapter 7	Dare To Dream	97

THE CURETTE

Chapter 1

Curettage Abortions, Uterine Perforation, Internal Bleeding & Trauma Experienced by Abortion Workers

---

It is fascinating to watch the development to maturity of a new human being by some technique such as sonar to which I have devoted much of my research life.

The observed miracle of healthy development is enhanced by recognising the cases in which it goes wrong, often recurrently so. It is particularly heart breaking to witness the present day wanton policy in many centres of discarding into the bucket or incinerator so much healthy unborn life whose only fault is that it is unwanted.[1]

Dr. Ian Donald

When I picked up the curette, the first thing I noticed was the weight. This instrument had some unexpected heft to it. The next observation I made was of the sensation of the cold steel. There is something about that sensation that

The Procedures

reminds you of the seriousness of this instrument. The one I picked up came as a set of three different sizes. Each one had a long handle designed for gripping. The handle had depressions along it, each one allowing its user to get a firmer grip. Beyond the handle was a rod which extended the scraping edge well beyond the handle so as to be able to use the scraping edge inside the uterus. Finally, I ran the scraping edge along my finger. To call it a scraper isn't entirely accurate. It's actually a blade that is a tad bit sharper than a butter knife. It's not sharp enough to cut my skin, but it is sharp enough to do some serious scraping. The blade reminds me of a teardrop extending from the rod. It is a rounded blade with both sides ending into the rod so that there are no sharp points. A sharp point inside the uterus could be disastrous.

The instrument I've just described for you is the Sharp Uterine Curette. The action of using a curette is called curettage.[2] A great variety of curettes are used throughout healthcare. What they all have in common is that they are all used to remove material by scraping. The curettes that most people recognize are the dental curettes used by dental hygienists to clean teeth. The hygienist will have an assortment of scrapers in different shapes that she uses to scrape the buildup of plaque off of teeth. The dental curette is different from the uterine curette in that the uterine curette has a distinctive blade for scraping.

The Curette

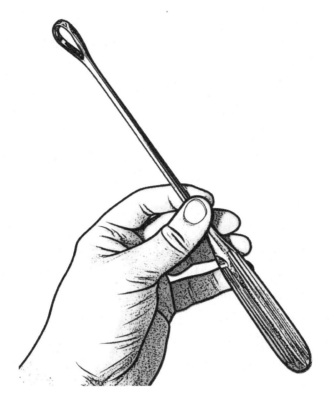

Sharp Uterine Currette

Before the introduction of the vacuum aspirator for abortions in the mid-twentieth century, the curette was the primary instrument used to bring about the death of the child and remove it. When I refer to the "primary instrument" in an abortion, I am referring to the instrument that is actually responsible for the death of the embryo or fetus. In order to understand what an abortion is, we have to begin with the curette.

The Procedures

When the curette is the primary instrument used, abortion doctors call it a "D&C" procedure. The term D&C refers to dilation of the cervix and curettage. The problem with this term is that it is so intentionally vague that you don't even know whether or not it is referring to an abortion. The D&C procedure is done for many legitimate health reasons other than abortion. Often it is used on women who are not pregnant to remove the lining of the uterus for unrelated health reasons. Further, a D&C could also be used to remove a fetus that miscarried. This is not an abortion because the D&C isn't causing the death of a fetus. The fetus has already died. The abortion industry engages heavily in vague terms like D&C because it conveniently distracts the patients, doctors, and staff away from the subject of the D&C abortion, the death of the unborn child.

A curettage abortion begins the way most surgical abortions do. The patient and the room have to be prepared for surgery. The cervix is numbed and manually dilated to allow the instruments to be inserted into the uterus. The preparation varies depending on the type of surgical abortion. I won't spend much time in this book on preparation as this isn't the controversial part of abortion.

With a curette abortion, the basic idea is simple. The curette is used to scrape throughout the uterus. In some cases the embryo may be scraped out intact. If

The Curette

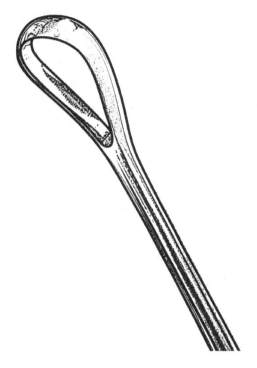

The Currette's Blade

this is the case, the embryo will die as a result of being separated from the life giving lining of the uterus and may survive for some time outside the uterus. The embryo or fetus is dependent on its mother through the placenta. The placenta allows oxygen and nutrients to transfer from the mother to the child. When the child is separated from the mother intact, it loses its source of oxygen and nutrients and dies relatively soon after. Everything else is scraped out as well, including the uterine lining, the placenta, and the umbilical cord. In most cases, the embryo or fetus does not come out

The Procedures

intact. The scraping can cause the fetus to be scraped out in pieces. The blade is used to separate the body of the fetus into smaller pieces to be scraped out of the uterus in a mass of blood, tissue, and fetal body parts. When this happens, then the death of the fetus is caused by the massive physical trauma of being dismembered.

The trauma may not be limited to the physical trauma experienced by the fetus. There may also be psychological trauma experienced by the abortionist and staff. Depending on how large and developed the fetus is, they will visibly see the body of the child that was aborted. If the abortion is early enough, you may not be able to distinguish the embryo from the bloody lining, sac, and placenta as it is scraped out. But a fetus in the latter half of the first trimester can be clearly seen as it is removed. As the blade is removed from the uterus, you may see a dismembered arm or leg wrapped around it. If you are especially unlucky, you may even see a beating heart or tiny face looking up at you as it is removed. In surgical abortions, it is very hard for the doctor and staff to avoid seeing the destruction they have caused to the unborn child's body.

The abortionist or staff has to take steps to ensure that everything has been removed. If something like an arm or leg is left in the uterus, it will become infected and can become deadly in a matter of days. To make

The Curette

sure everything is removed, the abortionist may hold everything up against a backlight so he can make out the various body parts and confirm that they are removed. In order to lessen the stigma of this violent act, the doctor or staff may number the body parts or use other tricks to avoid using the real names. So instead of saying, "We are missing the head", the staff member may say "We are missing number 1." This trick and many others like it are common in the abortion industry to dehumanize the child.

If the abortionist or his staff confront the fact that they are killing tiny humans, they may experience regret and quit doing abortions. An abortionist by the name of Dr. Anthony Levantino is a classic example of this. He appeared in a video on the website *www. abortionprocedures.com* saying:

> *In the early part of my career I performed over twelve hundred abortions. One day, after completing one of those abortions, I looked at the remains of a preborn child whose life I had ended, and all I could see was someone's son or daughter. I came to realize that killing a baby at any stage of pregnancy for any reason is wrong. I want you to know today, no matter where you're at or what you've done, you can change.* [3]

The Procedures

Another such example is that of Dr. Paul Jarrett who did abortions in the year following Roe v Wade. He had the following to say about his last abortion in which he had to switch from using a vacuum curette to a forceps due to the size of the fetus:

Initially, the abortion proceeded normally. The water broke, but then nothing more would come out. When I withdrew the [vacuum] curette, I saw that it was plugged up with the leg of the baby which had been torn off. I then changed techniques and used ring forceps to dismember the 13 or 14 week size baby. Inside the remains of the rib cage I found a tiny, beating heart. I was finally able to remove the head and looked squarely into the face of a human being - a human being that I had just killed. I turned to the scrub nurse standing next to me and said, "I'm sorry." I knew then that abortion was wrong and I couldn't be a part of it any longer. No one was critical of me for what I had done, nor for having stopped. But I had a lot of guilt about that abortion and had flashbacks to it from time to time. I sometimes dreamed about it. The guilt lasted about four years. [4]

The Curette

Notice how Dr. Jarrett talks about flashbacks and dreams about the abortion. Simply feeling guilty doesn't explain flashbacks and nightmares. But psychological trauma certainly could explain the flashbacks and nightmares. They could even be symptoms of post-traumatic stress resulting from the trauma of looking into the face of the aborted child. These kinds of testimonies resembling post-traumatic stress are not uncommon among doctors and staff who have left the abortion industry.

I can draw from my own experience to help explain post-traumatic stress and how it causes flashbacks. At one time I worked at a company where my job included operating a forklift. One day there was a tragic accident while I was operating the forklift. The result was a massive injury to a person's leg. The incident emotionally crippled me, but not because I felt any guilt. I knew it was an accident and that I wasn't to blame. As soon as first responders arrived I headed straight to the manager's office to try to regain my composure. One of my coworkers tried to assure me that it wasn't my fault. I responded that I knew it wasn't my fault, but it didn't make it any easier. For many days afterward I experienced flashbacks of the accident. At first I experienced them constantly. They tortured me. I relived the accident over and over again in head. I couldn't will myself to stop experiencing them. Eventually they became less frequent

The Procedures

until I stopped having them altogether. The reason I tell this story is to explain why I'd expect abortion workers to experience flashbacks. I only witnessed one mangled leg. Many of these workers have to witness all the mangled body parts: arms, legs, torso, and even the tortured faces. I don't see how anyone could not suffer traumatic stress from doing abortions.

We've looked at the physical trauma to the fetus done by the curette and the psychological trauma to the abortionist and staff. But we haven't looked yet at the psychological trauma to the mothers. We will look at this trauma further in chapter 12 regarding the abortion pill RU486.

There are a number of complications which can result from an abortion. Two common complications which can result in the death of the mother are uterine perforation and incomplete abortion. Uterine perforation is when the abortionist creates a hole in the uterus with an instrument like the curette. The abortionist can damage nearby organs. Uterine perforation can cause hemorrhaging or massive blood loss. In the most severe cases, the blood loss results in the death of the mother. Incomplete abortion is when parts from the prenatal human or her supporting structures are left in the uterus after an abortion. These parts can cause infection. In the most severe cases, the infection turns into sepsis, organ systems shut down, and the mother dies. In this chapter we will look at uterine perforation. We will look at incomplete abortion and sepsis in the next chapter.

Uterine perforation is when an instrument inserted into the uterus causes a hole in the wall of the uterus. All surgical abortions carry a risk of uterine perforation because all surgical abortions involve instruments in the uterus. It's important to understand that the uterus is a relatively fragile organ. You could think of the uterus like a stretchy bag. As the fetus grows, the bag stretches out to make room. When instruments like the curette are used inside the uterus, these instruments cause the walls of the uterus to stretch. What this means is that inserting instruments into the uterus has inherent risks. Perforations are relatively common in surgical abortions and are usually very minor and heal on their own. Many are so minor that they go undiagnosed. But this isn't always the case. Sometimes perforations are much more significant and won't heal on their own. If the perforation is too large, the patient may experience a large amount of blood loss. In the worst cases, the instrument may go through the wall of the uterus and damage surrounding organs such as the bladder or colon.

The Procedures

Very few abortion clinics are capable of treating a woman with a badly perforated uterus. Typically these women must be rushed to the emergency room for surgery. It is not uncommon for abortion clinics to call 911 and send their patients to the emergency room in an ambulance. At some of the least safe clinics, calling for an ambulance becomes a regular occurrence. A Planned Parenthood abortion clinic in St. Louis Missouri has become known for its regular emergencies, calling for an ambulance 74 times over ten years. It averaged an emergency every seven weeks.[4] If a woman with a significant perforation is not treated quickly, blood loss and infection could become very dangerous and even result in death. She could die from blood loss or from infection.

Dr. Paul Jarrett described a patient who came to him with a massive perforation that ultimately led to her death. This 18 year old young lady had flown to the State of New York for a legal elective first trimester abortion. At this time New York was one of only a few states that allowed elective abortions before Roe v. Wade. One of the arguments the pro-choice side frequently makes is that legalizing abortion will make it safer for the women. I can't emphasize enough that this was a legal abortion and that she was one of many women who lost their lives to legal abortions in the years leading up to Roe v Wade. He described the young woman's situation saying:

The Curette

When she returned home in terrible pain, she realized she was in trouble and for the first time, told her mother what had happened to her. Her mother contacted her own gynecologist, who in turn referred the patient to Coleman Hospital to be evaluated by the resident on call-me. Even though I was still wet behind the ears, I know that this pale, frightened little girl was still 10 weeks pregnant and her blood count was only half of what it should be. The private, attending doctor came in and took the patient to surgery immediately that night, where he repaired the hole that had been torn in the back of her uterus, which had caused her massive internal hemorrhage. Over the course of the next few days, infection set in which did not respond to antibiotics, and we made the painful decision to perform a hysterectomy. Tragically, the shock from the infection severely damaged her lungs and her course was steadily downhill. As I helplessly watched, she slipped into unconsciousness and a few days later she died.[6]

This story is of a woman who died from infection after her uterus was perforated. In some cases, women die from massive internal bleeding after a uterine perforation. That is what happened to a young woman

The Procedures

in my home state of Delaware. Gracealynn Harris was a nineteen year old African-American woman who already had a young son. She went to abortionist Dr. Mohammad Imran for a second trimester abortion at his abortion clinic in 1997. After the abortion, she was weak and left the clinic in a wheelchair. Before the end of the day she was dead from massive internal bleeding. No one understood how badly hurt she was until that evening when she had a seizure. The family sued Dr. Imran and his clinic and were awarded 2.2 million dollars. Dr. Imran was found to have committed malpractice in part because he did not use an ultrasound to guide his instruments. An ultrasound guided abortion is one where the ultrasound machine is used so the abortionist can see what he is doing with his instruments. Most abortions are not guided. You could call these blind abortions. Gracealynn Harris could still be alive today if her abortion had been ultrasound guided according to an expert witness.[7]

It is important to understand that not all abortions are equally risky. There are a number of factors that go into the quantity of risk that a woman is exposing herself to with an abortion. But unfortunately, the abortion industry isn't being honest with women and the public about the risks. Instead of allowing women to decide for themselves the amount of risk that they want to take, the abortion industry loudly proclaims that abortion is safe. And they have an army of

The Curette

journalists, politicians, and entertainers to push this propaganda for them. There is a certain amount of condescendence in this claim that abortion is safe. That's because safe is a subjective opinion based on the amount of risk you are comfortable taking. And so when the abortion industry says that abortion is safe, what they are really saying is that the risk of harming you is a risk the abortionist is willing to take. But what about the amount of risk that you are willing to take? Abortion isn't about the woman and her doctor making an informed decision together. This is an entire industry pushing their lack of risk aversion on America's women.

There are a number of red flags that women can look for to indicate that they are taking unnecessary risks. Here are some factors to look for in order to reduce the risk of complications.

Is the clinic very clean? If you see blood on the floor, insects, or anything dirty like that, you are looking at an unsafe clinic.

Does the doctor take time to build a patient doctor relationship? If the doctor doesn't take time to get to know you, your concerns, and your medical history, he may be an unsafe abortionist.

Are the staff and doctor rushed? Unfortunately, many clinics are having staff prepare patients for their abortions and you only see the doctor for ten minutes

The Procedures

when he comes in to do the abortion. A rushed doctor is an unsafe doctor.

Does the doctor do an ultrasound before the procedure? Doing an ultrasound beforehand is an incredibly important part of avoiding complications. What if you have twins? What if you have an ectopic pregnancy? There are so many things that can go wrong if they don't do a proper examination with an ultrasound.

The same is true of ultrasound guided and blind abortions. Does the doctor use an ultrasound to guide his instruments? Very few doctors do ultrasound guided abortions. The abortion industry considers ultrasound guided abortions to be optional and up to the personal preference of the abortionist. But if you want to avoid risk, you should demand it. Ultrasound guided abortions are far less risky than blind abortions.

How far along are you into your pregnancy? Abortions get much riskier as the fetus gets larger. This is especially true after the first trimester when the fetus is too large to be vacuumed out.

What is the doctor's record? Has he been disciplined by your state or local authorities? Has he lost any of his patients? What kinds of malpractice lawsuits has he faced and what were the results?

The Curette

Unfortunately, most women will not be asking these questions before they get an abortion. But as you can see, the amount of risk a woman takes with an abortion depends on a lot of factors. To make a blanket claim that all abortions are safe is irresponsible and a disservice to women. Women should be informed on a case by case basis the amount of risk involved and offered options like ultrasound guidance of instruments to reduce risk.

Finally, the abortion is not intended to be safe for the fetus or embryo. In this chapter we've looked at a specific type of abortion using a sharp uterine curette. The purpose of this instrument is the remove the embryo or fetus, usually in pieces. The blade of the curette is used to dismember the fetus and remove the parts: head, torso, and limbs. When other instruments are used to kill the fetus, the curette is often used to clean up and make sure nothing is left inside the uterus. Once you understand the curette and how it is used in abortion, you are now ready to learn about other types of abortion.

The Procedures

Chapter Notes

1. Willocks, James and Wallace Barr. *Ian Donald: A Memoir* London: RCOG Press, 2004. pp. 120

   https://books.google.com/books/about/Ian_Donald.html?id=6GtmNEpf2fwC

   Dr. Ian Donald (1910-1987) was an accomplished Scottish OBGYN, pro-life advocate, and inventor of the fetal diagnostic ultrasound.

2. Curette is pronounced [kyoo-ret].

   https://www.dictionary.com/browse/curette?s=t

   Curettage is pronounced [kyoo r-i-tahzh].

   https://www.dictionary.com/browse/curettage?s=t

3. "Aspiration" *Abortion Procedures: What You Need to Know*. Live Action

   https://www.abortionprocedures.com/aspiration/

4. Jarrett, Paul. "Testimony of Dr. Paul Jarrett, Former Abortion Provider" *Priests For Life*.

   http://www.priestsforlife.org/testimonies/1125-testimony-of-dr-paul-jarrett-former-abortion-provider

5. Downs, Rebecca. "Planned Parenthood: Women Need An Abortion Center That Can't Keep Them Safe" *The Federalist*. (2019)

   https://thefederalist.com/2019/06/03/planned-parenthood-insists-women-need-abortion-facility-ridden-medical-emergencies/

6. Jarrett

7. Vidmar, Neil and Valerie Hans. *American Juries* Amherst: Prometheus Books, 2007. pp. 83-87 https://books.google.com/books?id=jrhCTNdVgSAC&lpg=PA1&dq=american%20juries&pg=PA1#v=onepage&q=american%20juries&f=false

THE VACUUM

Chapter 2

Vacuum Aspiration Abortions, Incomplete Abortion, Infection, & Sepsis

Now that we've looked at the curette and how it is used in abortion, we can look at the most common surgical abortion today: vacuum aspiration. Vacuum aspiration is any abortion where the fetus/embryo, as well as everything else in the uterus, is sucked or vacuumed out. While vacuum aspiration continues to be the most common surgical abortion, the abortion pill is increasingly being used instead.

The primary instrument used in this abortion is a combination of two instruments: the curette, which is an edge for scraping, and a suction tube, which is called a cannula. It is often called a vacuum curette or a vacurette. You are already familiar with the sharp uterine curette from the previous chapter with its metal blade for scraping. In the vacuum curette, the scraping edge of the curette is incorporated into the suction tube.

The Procedures

Typical Disposible Vaccum Currette

Double-Sided Vaccum Currette

The Vacuum

The vacuum curette comes in many varieties. It comes in metal or plastic, straight or curved, rigid or semi-rigid, and reusable or disposable. It also comes in different sizes which are measured in millimeters. The style of vacuum curette used is left to the personal preference of the abortionist.

Vacuum aspiration can be further subdivided into aspiration with an electric pump versus manual vacuum aspiration which uses a hand-held syringe. Most vacuum abortions in the United States are done with the electric pump. In this abortion, the woman is prepared and the cervix is dilated and numbed as is necessary in surgical abortions. The doctor inserts the vacuum curette into the uterus and positions it. Then he turns on the electric pump in the aspirator. The vacuum curette is connected to the pump with plastic tubing. The electric pump is approximately ten times stronger than a household vacuum. It is very powerful and noisy. While the electric pump is turned on and creating a vacuum, the vacuum curette isn't engaged yet. The final step to begin the vacuuming is for the abortionist to engage a valve. Switching the valve applies the vacuum against the tubing and vacuum curette.

The Procedures

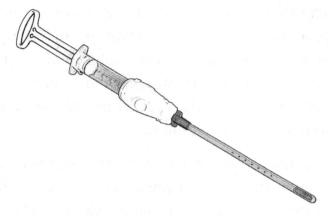

Manual Vacuum Aspirator

Manual vacuum aspiration uses a large hand held syringe. It is often referred to by its abbreviation, MVA. The vacuum curette is attached directly to the end of the syringe. MVA is nearly silent and the force of the suction isn't as strong. The type of vacuum used, electric or manual, is often left to the personal preference of the abortionist. While the electric pump is still the tool of choice for most abortionists, MVA appears to be growing in popularity both among abortionists and their patients.

Most abortionists use the electric pump because of the convenience and because they believe it to be less risky. The suction is more powerful which makes their job easier. In contrast, MVA requires significant arm strength to operate the syringe. MVA doesn't provide as much suction strength which means that

The Vacuum

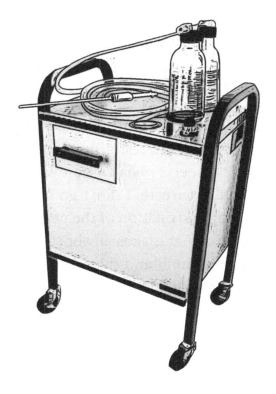

Electric Vacuum Aspirator

the doctor might have to use the syringe several times to get everything out. On the other hand, some doctors do prefer MVA over the electric pump because it is mostly quiet. The sound of the electric pump can be disturbing for the mothers, especially if the mother understands what the pump is doing to her fetus. MVA also allows the abortionist to have more direct control over the suction. Abortionists disagree over whether or not MVA is more likely to cause complications. It

The Procedures

may require the cervix to be dilated more and it may require the abortionist to spend more time in the uterus, both of which would increase the risk of complications.

A vacuum abortion begins like any other surgical abortion with the preparation, numbing, and dilation of the cervix. The cervix has to be dilated enough to insert the vacuum curette. I don't go into great detail into the numbing and dilation of the cervix in this book because that isn't what makes an abortion an abortion. If all you did was numb and dilate the cervix, no one would care about abortion. There would be no pro-life movement if all we did was operate on the cervix. The embryo/fetus is the focus of the abortion. The abortion industry, on the other hand, focuses heavily on the cervix. They see it as a convenient distraction away from the embryo/fetus. That's why they name their procedures names like dilation & curettage (D&C) and dilation & evacuation (D&E). The goal is to keep you focused on the dilation of the cervix and away from the violent end of your embryo.

After dilation, a typical vacuum aspiration abortion begins with the abortionist inserting the vacuum curette into the uterus. The abortionist positions the curette where he thinks it should be. In the rare instance that an ultrasound is used to guide the

The Vacuum

abortion, the curette is placed exactly where it needs to be to get the embryo/fetus. The tubing from the electric vacuum aspirator then slides over the other end of the vacuum curette. The tubing goes back to the aspirator which has two glass jars where the fetal remains are collected. The aspirator is then turned on. The pump makes a creepy sort of humming and thumping sound. It's particularly disturbing because we know what it is about to do. If a woman suffers from post-traumatic stress from her abortion, the memory of the sound of the aspirator may be attached to her trauma. Similar sounds like the sound of a household vacuum turning on may startle her or cause flashbacks.

Now the pump is turned on but the vacuuming hasn't begun yet. When the abortionist is ready, he has to engage a valve on the aspirator that directs the suction against the tubing and curette. The bright red blood from the lining of the uterus can be seen through the tubing and into the jars. If he damages the uterus, the blood from the bleeding uterus is vacuumed as well. In most instances an ultrasound is not used to guide the abortionist. He moves the vacuum curette back and forth in a twisting motion to blindly scrape the inside of the uterus. If it is ultrasound guided, he can move the vacuum curette to get the fetus without blindly scraping.

The Procedures

If he fails to make contact with the embryo/fetus, the abortion is called a failed abortion because the fetus continues to live and grow. The woman is still pregnant. When the abortionist does make contact, the force of the suction removes the embryo in pieces. The arms, legs, torso, and head are all suctioned out piece by piece through the tubing and into the jars. Also suctioned out are the supporting structures: the umbilical cord, placenta, and membranes. Sometimes the head presents a problem for the abortionist. It is the hardest part to remove due to being large and round. If the child is too large, the head will not be suctioned out. The abortionist has to use a forceps and sharp uterine curette to crush it and remove it. Further, if the fetus is too large, other body parts such as a leg or arm may get stuck in the end of the vacuum curette and force the abortionist to switch to a forceps abortion.

The contents of the jars are then taken to be examined. Usually this is done in a separate room designed specifically for the purpose of examining the remains. Some clinics call this the "POC room," which stands for products of conception. POC is just one of the many ways abortionists and their staff dehumanize the child so as not to face the reality of what or who they just aborted. A more accurate name would be

The Vacuum

something like the fetus examination room. At this point someone must examine the remains to make sure that everything was removed. The abortionist or staff member holds the remains against a back light in order to see everything. Did they get the head? Did they get two legs and two arms? Did they get the rib cage? Staff members may also invent dehumanizing code language to describe the parts. For example, an arm might be described as #2. So if they only have one arm, they may tell the abortionist that they are missing a #2. The amount of destruction to the fetal remains differs between electric and manual vacuums. The electric vacuum is stronger. The body is likely to be dismembered into a larger number of pieces. Some of those pieces may be destroyed beyond recognition. MVA, on the other hand, is more likely to result in an intact embryo or fetus being removed. The weaker suction is less destructive to the body.

When an abortionist misses a fetal part or supporting structure and leaves it in the uterus, it is called an incomplete abortion. In an incomplete abortion, the fetus is successfully killed. This is in contrast to a failed abortion where the fetus survives and continues to grow. Incomplete abortion is one of the more common abortion complications, along with uterine perforation and hemorrhaging.[1] Incomplete

The Procedures

abortion is made worse by the fact that most abortionists do the abortion blind, not ultrasound guided. When done blind, the abortionist cannot see if he got everything until it is examined against the backlight. But even with the backlight, it isn't hard to miss something. It all comes out as a mass of tissue and blood. While the untrained eye can identify an arm or a leg, it takes a trained person to identify everything and make sure that it is all removed. According to *Management of Unintended and Abnormal Pregnancy*, "Failed attempted abortion is usually recognized by immediate gross tissue examination with backlighting or the use of magnification when necessary. Clinic staff members can be trained to become proficient examiners of tissue specimens."[2]

An incomplete abortion has the potential to become dangerous very quickly. This is especially true if the uterus has become injured in addition to the incomplete abortion. The abortionist's instruments can injure the uterus. But the left over pieces from an incomplete abortion can injure the uterus as well. For example, a sharp bone fragment left in the uterus can cut into the side of the uterus. An incomplete abortion, especially when coupled with uterine injury, can become infected. A women with an infection from abortion needs to be treated right away. If she is not treated, the infection can turn into sepsis.[3] Sepsis is when the body is extreme in its response to infection. The body starts a chain reaction that can result in tissue damage and entire organ systems shutting down. Ultimately, sepsis will result in death.[4] That is why it is so important that abortion patients quickly get treatment if they suspect infection or sepsis. What started as a routine abortion can turn deadly in only a few days if left untreated.

One of the more well-known cases of a woman dying from sepsis from an abortion is the case of Marla Cardamone. She was an 18 year old from Allegeheny County Pennsylvania who went to the prestigious Magee Women's Hospital on August 15, 1989. She was there under pressure from a social worker for a late-term abortion. The social worker had erroneously convinced her that her baby would have birth defects from the medication she was taking. She received a urea installation abortion which involves killing the fetus with urea before inducing labor to deliver the dead baby. This type of procedure is explained in chapter 16. Unfortunately for Marla, the labor did not go as planned. Instead of delivering her dead child, she became badly infected. The infection became sepsis. As her body was shutting down, she suffered from violent seizures and vomiting which wrecked her body.

The Procedures

In only a matter of hours she was dead. Marla's mother had the following to say about her daughter's death.

> *Finally, they allowed me to see Marla's body. When I entered the room, I could hardly believe what I saw. There was my beautiful daughter so horribly disfigured that she was almost unrecognizable. A tube was still protruding from her mouth and I could see that her teeth and gums were covered with blood. Her eyes were half opened and the whites of her eyes were a dark yellow. Her face was swollen and discolored a deep purple. The left side of her face looked like she had suffered a stroke. All I wanted was to hold her. I managed to get an arm around her and kissed her good-bye.*

Marla's story didn't end like so many other women who have died from legal abortions. That's because her family decided to do something bold. They released the autopsy photographs to the public. The pictures were chilling. Her body was bruised and swollen. Her teeth and mouth were caked with blood. But the most disturbing picture was the one where the person doing the autopsy opened her uterus and showed her dead child still inside her. As if to add insult to injury, the

The Vacuum

hospital did not release the child's body to the family for a proper burial. Instead it was disposed of as medical waste.[5] The autopsy photographs can be seen at www. safeandlegal.com.

So far we have looked at curette and vacuum abortions as well as the emotional burden on abortion staff and the risk of incomplete abortion, infection, and sepsis. In the next chapter we will be looking at the abortion pill and the emotional trauma experienced by abortion patients, especially from the abortion pill.

The Procedures

Chapter Notes:

1. Paul, Maureen, et al. *Management of Unintended and Abnormal Pregnancy* Hoboken: Wiley-Blackwell, 2009, pp. 228

2. Ibid, pp. 227

3. Ibid, pg. 228

4. "What is sepsis?" *Center for Disease Control and Prevention.* (2019)

   https://www.cdc.gov/sepsis/what-is-sepsis.html

5. Dunigan, Christina. "Legal Abortion Death: Marla Cardamone, 18" *Clinic Quotes.* (2012)

   https://clinicquotes.com/legal-abortion-death-marla-cardamone-18/

THE ABORTION PILL RU486

Chapter 3

Mifepristone Abortions & Trauma Women Experience

I grabbed a towel to bite on in order to keep from screaming and was nearly passing out. As I got up I saw blood everywhere. I saw parts of my baby, images I will never be able to erase. I fell to my knees in pain and was blacking out. Concerned that the guys would see all the blood and clumps, I got on my knees and cleaned it up.

Ann[1]

In this quote, Ann is writing about her experience with the abortion pill, also known as RU486, mifepristone, or medical abortion. RU486 is the brand name for mifepristone. Ann's story is not unique. I've read countless stories from women just like her. Unlike surgical abortions, the abortion pill causes the abortion while the woman is alone at home or at a hotel room. Most surgical abortions are relatively quick. You simply go to the clinic and let the abortionist do the

The Procedures

hard part. But with the abortion pill, there is no abortionist there to do the hard part. You have to clean it up yourself. Abortion clinics instruct women to have the abortion on a toilet bowl and flush. This has caused some people to start referring to these abortions as toilet boil abortions.

In order for the abortion pill to be successful, everything must be passed out of the uterus. We are going to look more at Ann's story and the downsides to the abortion pill that abortion clinics rarely warn about. But first let's look at how the pill causes an abortion.

An abortion with pills actually requires two different types of pills. The first pill is mifepristone, which most people know by the brand name RU486. This is the pill that blocks progesterone and causes the death of the embryo. The second pill is Misoprostol. This pill is taken 24-48 hours after to cause contractions to expel everything out of the uterus.

To understand how RU486 kills the embryo, it is important to remember some of the biology from chapter 5 where we learned about the role the corpus luteum plays in producing progesterone. It is also important to remember chapter 8 where we learned that soon after implantation the blood of the embryo and the blood of the mother are able to exchange nutrients in order to grow and support the embryo.

This is done when the embryo's blood in early placenta tissue called the trophoblast comes in close contact with the mother's blood in the lining of the uterus. We also know that the follicle that ovulated the egg turns into a gland called the corpus luteum which produces progesterone. The word progesterone simply means "pro-gestation" and is key to gestating the embryo.[2] The progesterone thickens the lining of the uterus and keeps that lining nice and healthy to support the embryo. In order for the lining to benefit from the progesterone, there are receptors in the uterus that must receive that progesterone. The embryo in return produces a hormone called hCG which causes the corpus luteum to continue producing progesterone. Without an embryo and hCG, the progesterone stops. The lining deteriorates or breaks down and is expelled as menstrual blood.

RU486 interrupts the process. This drug essentially tricks the progesterone receptors in the uterus. The receptors receive the RU486 as if it were progesterone. But the RU486 doesn't activate the receptors to support the lining of the uterus. The progesterone can't be received because the RU486 has already taken up those receptors. This is why RU486 is called a progesterone blocker. It literally blocks the progesterone from the receptors and renders the receptors useless. Without

The Procedures

the progesterone, the lining begins to deteriorate and break down the same as it would if the mother were not pregnant. Only this time she is pregnant.

What this means is that the embryo's blood is no longer able to exchange nutrients and gasses with the mother's blood. As the cells of the lining die off, the placenta tissue of the embryo is separated from the lining of the uterus.[3] The blood of the embryo and the blood of the mother are no longer in close contact. The embryo can't expel waste and it can't receive nutrients and oxygen. This is what causes the death of the embryo. In essence, the embryo dies of neglect. Neglect in a parent-child relationship is when the parent fails to provide for the basic needs of the child. Failing to feed a child a minimally nutritious diet, failing to provide a safe living environment, and failing to clothe a child are examples of neglect. In an RU486 abortion, the mother fails to provide for her embryo the necessary nutrition and also fails to provide the necessary environment in the uterus.

It is important to understand that death through neglect is still an act of violence. Some people see RU486 abortions as less morally objectionable. Many pro-life people seem less motivated to end RU486 abortions than surgical abortions. But killing is always a violent act regardless of how the killing occurred. I think some of the confusion in this area is due to a misunderstanding of types of violence. Killing is an act of violence but so is significant bodily harm. Some violent acts are killing. Some violent acts are not killing but are acts of bodily harm, such as breaking an arm or even giving someone a black eye. Many violent acts are both. This is true of surgical abortions. The further developed the fetus becomes, the greater the acts of bodily harm that are done in the process of killing that fetus. And so late-term second and third trimester abortions are rightfully seen as the most violent. For example, the crushing of the skull is not typically done in the first trimester. But it is done later in pregnancy. The act of crushing the skull is increasingly violent. RU486 abortions are not doing direct physical trauma to the body of the embryo in order to kill it. But these abortions are still causing the intentional death of the embryo, which necessarily means that all abortions are acts of violence, even the very early ones. It is important to understand that every abortion necessarily kills a prenatal human. If abortion didn't kill, it wouldn't be an abortion. If abortion didn't kill, there would be no pro-life movement. This is why abortion is fundamentally different from any other medical practice.

Once the embryo has died, its body, along with the uterine lining, deteriorates, making it easier to expel from the uterus. This is the reason RU486 is used. It is not necessary to block progesterone in order to expel

The Procedures

the embryo from the uterus. Some abortionists have been known to skip the RU486 entirely and simply induce labor to deliver the embryo. But the abortion industry has found that it is much easier to expel the embryo after its body, along with the uterine lining, has deteriorated. It requires weaker contractions to expel and there are fewer complications.

The second pill used in RU486 abortions is Misoprostol. This is the pill used to cause cramping, contractions, and labor. When a woman has her period, she naturally has very small contractions or cramping. RU486 causes the uterus to attempt a normal period including the cramping. It is not a normal period, however, because the woman must pass the dead embryo. And it is not like a normal period because the Misoprostol induces contractions that are stronger than a normal period. Misoprostol is also used in labor and delivery wards as another way to induce labor. It is known in labor and delivery for causing very strong contractions. Some even describe the contractions as violent. The important thing to remember is that it induces labor. It is not just like a heavy period. And it is not like having an early miscarriage, unless your doctor gives you something like Misoprostol to help you pass the miscarried child. This is important to understand because abortion clinics frequently downplay and mislead women about the amount of pain and blood they will experience. If a woman knows how much pain

The Abortion Pill RU486

and difficulty she is signing up for, she may choose a surgical abortion instead or she may even choose not to have an abortion. Planned Parenthood describes it on their website as "like having a really heavy, crampy period" and "very similar to an early miscarriage."[4]

Ann's story is typical of the countless stories I've read about women having RU486 abortions with Misoprostol. She describes the Misoprostol as follows:

> *It was time for me to take the last dose. As I put the pills in my mouth and let them dissolve, within ten minutes I started to feel intense cramps. When the cramps became unbearable I made my way to the bathroom. I locked the door and experienced the most severe pain I had ever felt in my life. I sat on the toilet and bent over in pain. I wanted to scream but my ex-boyfriend and his friends were right outside the door in the living room watching TV. It was a small apartment. I grabbed a towel to bite on in order to keep from screaming and was nearly passing out. As I got up, I saw blood everywhere. I saw parts of my baby, images I will never be able to erase. I fell to my knees in pain and was blacking out. Concerned that the guys would see all the blood and clumps, I got on my knees and cleaned it up. As soon as I left*

The Procedures

> *the bathroom I was about to faint when my ex-boyfriend helped me to bed.*"[5]

The pain and blood can be very traumatic and unexpected but are not the only traumatic factors. Seeing the body parts of the embryo after it has passed can also be traumatic. Abortion clinics don't typically warn women about what they will see although a few clinics may urge women to flush without looking. RU486 abortions can be done up to ten weeks gestation. This is the point when we begin calling it a fetus due to the fact that it clearly looks like a baby. When having this abortion, women can see clearly the features of the child, including fingers, toes, face, umbilical cord, and placenta. Sometimes the child comes out intact. Sometimes she comes out in pieces. The child may or may not be inside the gestational sac. You will remember that embryonic folding occurs by the sixth week. The child is about the size of a pea at six weeks and as large as a strawberry by ten weeks. As you can imagine, passing a strawberry-sized body through the cervix would take significant contractions. Picking that child up and placing it in the toilet to flush it must create a heavy emotional load.

It's important to mention the men as well. Abortion certainly has an emotional impact on the fathers. These men are often the most forgotten in the abortion issue. Some men want their babies and suffer greatly when they find out that their children were aborted and they were powerless to stop it. Other men carry the guilt of coercing her, manipulating her, or paying for her abortion. Abortion has a traumatic effect on all involved: the abortionist, the mother, the father, and even the rest of the family.

In 2016 the FDA approved significant changes to the protocol for RU486 abortions, which had the effect of expanding RU486 abortions and making them more traumatic for women. The change did two things. First, it increased when you can get this abortion from seven weeks gestation to ten weeks gestation. Second, it decreased the dosage of RU486 and increased the dosage of misoprostol.[6] By increasing the gestation to ten weeks, the FDA dramatically increased the number of abortions that could be done with the pill. It also means that women are passing significantly larger fetal bodies. At ten weeks, we are talking about a strawberry-sized body. The body parts are much larger and more distinct. The woman likely will see the body, possibly intact or possibly in pieces.

The change in dosage also increases the trauma for women. The RU486 dosage was decreased from 600 milligrams to 200 milligrams, a third of the dose. This was a big win for abortion clinics because RU486 is an expensive drug. A third of the dose equals a third of the cost for the abortion clinic to buy that drug. It

The Procedures

also means that the lower dose is less successful in killing the embryo. If a woman only takes the RU486 and doesn't finish with the misoprostol, there is a significant chance the embryo could survive and continue with no birth defects. The dosage of misoprostol increased from 400 milligrams to 800 milligrams with the possibility of taking additional misoprostol. And so now the chances of the embryo being loosened from the lining are less and the contractions needed to expel the embryo are much greater. Doubling the dosage of misoprostol means stronger contractions and more pain. All of these factors are combined to make RU486 a more traumatic experience for women.

Some of the more common side effects of RU486 are nausea, vomiting, diarrhea, and headaches. Sometimes heavy bleeding can last for weeks.[7] The more severe complications include what is called an incomplete abortion. Incomplete means that parts of the child or supporting parts are still in the uterus. These parts will result in an infection which can become very dangerous and even life-threatening. If a woman doesn't pass everything, then she must go back for a surgical abortion to finish the job. The story of Ann which we looked at earlier in this chapter was an incomplete abortion. She commented, "As they performed the D&C I couldn't help but think my baby

42

was a fighter."[8] Another serious complication is an ectopic pregnancy. RU486 abortions only work if the embryo is in the uterus. An RU486 abortion on an ectopic pregnancy is dangerous. Some other more serious complications include interactions with other drugs, allergic reaction to the drugs, and the presence of an IUD. RU486 can in some rare cases cause death. According to the FDA, 24 women have died from RU486 complications in the United States between September of 2000 and the end of 2018.[9] While abortion supporters emphasize that these deaths are rare, the deaths are real. About one or two women typically die in the United States each year from RU486.

The FDA has regulations in place to try to reduce the risk of complications and death. These regulations are called REMS, which stands for Risk Evaluation and Mitigation Strategy. The FDA requires that doctors be certified in the REMS program in order to dispense the drugs. In other words, not just any doctor can write a prescription for RU486. The doctor has to dispense the RU486 in a clinic or hospital setting. You can't get this drug at your local pharmacy. In order to get certified, the doctor has to meet minimum competencies, such as the ability to do surgical abortions and diagnose ectopic pregnancies. The doctor must also have access to certain equipment, such as the equipment necessary to provide blood transfusions.[10]

The Procedures

Not surprisingly, the abortion advocates are fighting hard to do away with REMS and any other restrictions designed to protect women. The ultimate goal is to make RU486 as easy to get as possible, including over the counter. As an over the counter drug, there would be no supervision by a doctor. None of the precautions in REMS would be in place. The abortion industry is trying to expand RU486 in some other ways on the road to over the counter abortions. One that many people have heard of is the so-called webcam abortions which are done via telemedicine. Planned Parenthood of Iowa first rolled out a model where the patient would go to an abortion clinic, but the doctor would not be on site. Instead the doctor would interact with the patient and clinic staff over webcam. Many states have now banned this practice but a few clinics still do these webcam abortions. The latest FDA approved trial that is now being conducted is an at-home webcam abortion. In this trial, the interaction with the doctor would be in your own home and the pills would be mailed to you.

Webcam abortions are ultimately a step toward the eventual goal of over the counter abortions. Abortion advocates call this "self-managed abortion." Ironically, Roe v. Wade was supposed to get us away from dangerous "self-managed abortions." After all, isn't this why pro-choice activists are running around

The Abortion Pill RU486

waving coat hangers? But with RU486, self-managed abortions are making a comeback. RU486 has drastically changed the abortion landscape. Regardless of the fact that the FDA prohibits self-administered RU486 abortions, some women still do them. It is surprisingly easy to buy black market pills. You don't even have to go to the dark web or a sleazy drug dealer to buy them, although some people may do that. People buy black market pills online. It's not safe and the FDA warns people not to do it. But people do it anyway. This is likely the future of illegal abortions. It is already happening. When Roe v. Wade is overturned and some states prohibit most abortions, RU486 will likely be the illegal abortion of choice. But this doesn't stop pro-choice activists from waving around coat hangers and making themselves look silly.

The abortion pill has drastically changed the way we do abortions in the United States and around the world. It has allowed women to see the humanity of their aborted children in a way that they couldn't with surgical abortion. It has made the abortion procedure longer, more painful, and more traumatic. But one last change that we must look at is the way it has made abortion more profitable. RU486 has been a gift to the abortion industry. Not only can they typically charge more for RU486 than for a surgical abortion, they can

The Procedures

save on a lot of the costs as well. There is less equipment and overhead involved. The abortion itself only requires the woman to take a pill before she walks out of the clinic. The clinic doesn't have to prepare the woman for surgery. The doctor doesn't have to take the time to do the surgery. It's super easy. In fact, the overhead for RU486 is so much lower that there are abortion clinics all over the country that only do RU486. It's much easier and less expensive to open a new abortion clinic if it only offers RU486.

Another benefit to the abortion industry is that the abortionist doesn't have to deal with the complications should they arise. The complications from surgical abortions can be very costly for the abortionist and the clinic. An abortionist may have to deal with complications like uterine perforation and hemorrhaging himself in the clinic. If he has to send the woman to an emergency room, he would have to face the negative publicity of an ambulance at his clinic. He may also face the risk of medical malpractice lawsuits from trying to treat complications in the clinic. With RU486, the clinic can direct the woman to go to the emergency room if necessary and let her complications be some other doctor's problem. There is no ambulance at the clinic. There is no 911 call from the clinic. There is no bloody mess to clean up in the clinic. It is a really great benefit for the abortion clinic.

The Abortion Pill RU486

The great losers in RU486 abortions are the unborn children and their mothers. I can't imagine the horrific experience of being alone, experiencing contractions, passing my perfectly formed little child, and flushing that toilet. These women deserve to at least know the truth before they decide to do this abortion. They deserve all the details, including how far the child is developed, the severity of the contractions, and what they might see and experience. At a minimum, they deserve the truth.

The Procedures

Chapter Notes

1. "Ann" Taken from Facebook on March 27, 2019. Ann's real name has been withheld to protect her privacy. Lightly edited for grammar and readability.

2. Paul, Maureen, et al. *Management of Unintended and Abnormal Pregnancy* Hoboken: Wiley-Blackwell, 2009, pp. 113

3. Ibid, pp. 113

4. "The Abortion Pill" *Planned Parenthood.*

   https://www.plannedparenthood.org/learn/abortion/the-abortion-pill

5. "Ann"

6. Rafie, Sally. "Abortion Pill Label Change: What Pharmacists Need to Know" *Pharmacy Times.* (2016)

   https://www.pharmacytimes.com/contributor/sally-rafie-pharmd/2016/04/abortion-pill-label-change-what-pharmacists-need-to-know

7. "Medical Abortion" *Mayo Clinic.* (2018)

   https://www.mayoclinic.org/tests-procedures/medical-abortion/about/pac-20394687

8. "Ann"

9. "Mifepristone U.S. Post-Marketing Adverse Events Summary through 12/31/2018" *U.S. Food and Drug Administration.*

   https://www.fda.gov/media/112118/download

10. "Prescriber Agreement Form: Mifepristone Tablets, 200 mg" *U.S. Food and Drug Aministration*

    https://www.accessdata.fda.gov/drugsatfda_docs/rems/Mifepristone_2019_04_11_Prescriber_Agreement_Form_for_GenBioPro_Inc.pdf

ABORTION PILL REVERSAL

Chapter 4

Progesterone Support to Reduce the Effects of RU486

"I wanted so badly to have this baby and to have a second chance."

Rebekah Buell-Hagan[1]

After RU486 was approved by the FDA for abortion in 2000, something interesting happened. Woman started coming forward after taking the first pill and changing their minds about going through with the abortion. Sometimes people make impulsive decisions and end up regretting those decisions. This is especially true when a woman can walk into an abortion clinic and in almost no time at all, ingest a pill to cause an abortion. She may impulsively go to an abortion clinic when she first discovers that she is pregnant and emotions are running high. But after she takes the first pill at the clinic, she may put some more thought into it and realize that this isn't the decision she wants to make. This is exactly what happened in 2007 when a lady in North Carolina approached her family doctor after changing her mind. Her doctor, Dr. Matthew Harrison, did some research on how RU486 blocks progesterone and decided to try an idea. He injected her with progesterone that he happened to have in his office. He knew it was a long shot and warned her that it may not help. Six months later she became the first recorded instance of a woman delivering a healthy baby after taking RU486 and then progesterone in an attempt to save the child.[2]

On the other side of the country, another doctor named Dr. George Delgado had a similar experience giving a patient progesterone and saving her child. As word spread of this case, Dr. Delgado began receiving more and more calls from people wanting to save their unborn children after taking RU486. Dr. Delgado, Dr. Harrison, and another provider named Dr. Mary Davenport worked together to start a nationwide program to administer progesterone to women who regretted their decision to take RU486.[3] Today that program is called Abortion Pill Rescue and has a nationwide network of hundreds of providers that prescribe and administer progesterone. A nationwide network of pro-life pregnancy care centers administers a 24/7 hotline to connect women with providers. You

Abortion Pill Reversal

can learn more about the network at www. AbortionPillReversal.com.

The idea of administering progesterone for pregnancy is not a new one. For decades fertility doctors have been giving progesterone regimens to patients to help with fertility and reduce the risk of miscarriage. Anyone who has gone to a fertility specialist is likely to be familiar with progesterone support. It is one of the more memorable treatments as it usually includes a painful injection into the muscle. Progesterone is typically administered three different ways: an intramuscular injection, orally, and with a suppository. The progesterone suppository is like a large pill that is inserted into the vagina. A doctor may prescribe a combination of the three to get the desired dosage. While the term "progesterone support" is used in fertility clinics, doctors trying to reverse an RU486 abortion use the term "abortion reversal." But these terms both refer to administering progesterone in order to prevent the loss of the embryo or fetus. I prefer the term "progesterone support" as it is more descriptive of the specific medical care that is being provided.

It shouldn't be surprising that progesterone support to undo RU486 is controversial. The abortion industry is quite vocal with their opposition to progesterone support. If a woman changes her mind

The Procedures

and calls the abortion clinic, she will typically be told that she has to take the second pill and finish the abortion. Women are strongly discouraged from attempting progesterone support. Some abortion clinics will even go so far as to falsely claim that RU486 will give the baby birth defects. [4,5] This is despite the fact that there is zero evidence that RU486 causes any kind of defects or deformities. Even the American College of Obstetrics and Gynecology, which is vocally pro-choice and vocally opposed to abortion pill reversal, in a 2017 paper, said that RU486 "is not known to cause birth defects."[6]

But the most widely made and unfounded claim by critics is that progesterone has too many possible side effects and is dangerous to women. But progesterone support is widely accepted as extremely safe and is a common treatment both for infertility and for women going through menopause. It is true that any medication can carry risk. But the risks associated with progesterone support aren't substantial. And yet this is a common claim that is unfortunately used to scare women.

Those critics of progesterone support who are more objective and honest about the medicine will tell a woman who has changed her mind that there is some chance her baby could survive. Even without progesterone support, there are two possible steps a woman can take to give her embryo/fetus a chance at

Abortion Pill Reversal

survival. If it has been less than an hour after she took the first pill, she could induce vomiting or have her doctor induce vomiting to attempt removal of the RU486 before all of it has been digested. A woman would have to act fast for this to be possible. The other step is to simply refuse to take the second pill and hope for the best. There is a lot of disagreement as to the chances of an embryo surviving only the first pill. But the chance of survival could be as high as 50%.[7] It is also noteworthy that in 2016 the FDA approved a significantly lower dose of RU486. It is now recommended that the first pill have 200 milligrams instead of 600 milligrams. The lower dose could mean a greater chance of survival if the woman doesn't take the second pill.[8] Most women, however, are not informed that these two options exist. If she were to call her abortion provider and express a desire to continue the pregnancy, it is not likely that she would be informed of the possibility of continuing. It is more likely she would be told that she must take the second pill and finish the abortion.

Most of the controversy surrounding progesterone support to reverse RU486 is over whether or not the progesterone increases the likelihood of a woman being able to save her baby. The critics, who reside mostly in the abortion industry, argue that added progesterone does not increase the patient's chances of continuing the pregnancy. They further claim that doctors doing it are giving women false hope. The doctors doing the progesterone support claim that it does increase the patient's chances of continuing the pregnancy and that the abortion industry isn't giving the patient enough hope that it can be done.

In 2018, Dr. Delgado and his network published a case study showing the results of their work. It analyzed the results of 547 women over four years and using different types of progesterone support. He calculated the survival rate and published the results. He found an overall survival rate of 48%, meaning that 48% of the women using progesterone support were able to carry their babies to live birth. He contends that the survival rate would have only been 25% without the progesterone according to previous research. And so he is claiming that the progesterone significantly increases the chances of live birth. The critics are now claiming that they don't accept a 25% survival rate without progesterone. They now think its 50% and therefore the progesterone didn't help. In my opinion, the critics are just moving the goal posts. What's interesting about the study is that the thirty eight women receiving the highest amount of progesterone, six or more intramuscular injections, were the women most likely to have a live birth. They had a survival rate of over 89%.[9] And so how does the chance of survival increasing

Abortion Pill Reversal

with higher dosages if the progesterone support doesn't work? If the progesterone support doesn't work, we would expect about the same survival rate regardless of the dosage. But if the progesterone support does work, then it makes sense that greater support would mean higher survival rates. In my opinion, the higher survival rate is compelling evidence.

It is important to understand the difference between a case study and a randomized trial. A case study is simply reporting on the results of the treatment. In other words, it only reports what the doctor did for his patients and the results. Published case studies are evidence and are worthwhile. But randomized trials are far more compelling. A randomized trial is much more rigorous and compares those who received the treatment to those who received a placebo or didn't receive treatment. Dr. Delgado has acknowledged that a randomized trial is needed. But he won't do the trial because it would be unethical to give a placebo to a woman who wants to save her baby.[10]

As of the writing of this book, an abortionist named Dr. Mitchell Creinin is attempting a small randomized placebo-controlled trial that will include forty women. Presumably the women he will recruit are patients at his abortion clinic. His ethical justification is that all of these women are seeking abortions anyway. He will give some of these women progesterone and others a

The Procedures

placebo in order to analyze how many of their embryos are still alive two weeks after they took the RU486. After the two weeks are up, he will do surgical abortions on any remaining living embryos and fetuses.[11] And so he will be experimenting on these embryos and possibly saving some of these embryo/fetuses with progesterone, only to surgically abort them after they no longer benefit his research.

It's important to understand the background of Dr. Creinin and the organization funding his study. Creinin is an abortionist's abortionist. He is widely recognized as an expert. He was an expert witness when congress passed the Partial Birth Abortion Ban and is a co-author of *Management of Unintended and Abnormal Pregnancy*, the primary textbook used to train abortionists. He is also known to have been cited by the FDA for violations in a previous study.[12] Most notably, Creinin is already a vocal critic of abortion pill reversal which begs the question of whether or not he can do the research objectively and not allow his biases to skew the research.[13] A further problem for this study is the organization paying for it. Society of Family Planning is a pro-choice organization seeking to expand abortion, especially RU486 abortions. This organization spends millions of dollars each year in research grants with the bulk of the money going toward expanding RU486 as widely as possible. In 2018,

Abortion Pill Reversal

over eight million dollars in grants were awarded to expand RU486 in areas such as mail-order abortions, self-administered abortions, over the counter abortions, and abortions in family practice.[14] All this is to say that this small randomized trial is being conducted with an agenda. Not only must Dr. Crenin do the research, but he must also be able to convince the public that his research is legitimate.

Editors Note: The Crenin study was cancelled and annouced after the publication of this book.

The Procedures

Chapter Notes

1. "One Young Woman's Abortion Pill Reversal Story" EWTN. (2018)

   https://www.youtube.com/watch?v=TWyf6Y_BPlQ

   Quote taken from interview at 2 minutes 30 seconds Ms. Buell-Hagan was a successful abortion pill reversal patient and is now an advocate for reversal.

2. Cleveland, Margot. "Are Abortion Reversals Science or Scam?" *The Federalist.* (2017)

   http://thefederalist.com/2017/05/16/abortion-reversals-science-scam/

3. Ibid

4. Devine, Daniel. "Cynthia's choice" *World Magazine.* (2013)

   https://world.wng.org/2013/04/cynthias_choice

5. Hobbs, Jay. "On Abortion Pill Reversal, It's Time to Hear from the Women" *Pregnancy Help News.* (2018)

   https://pregnancyhelpnews.com/on-abortion-pill-reversal-it-s-time-to-hear-from-the-women

6. "Facts Are Important: Medication Abortion "Reversal" Is Not Supported By Science" *The American Congress of Obstetricians and Gynecologists.* (2017)

   https://www.acog.org/-/media/Departments/Government-Relations-and-Outreach/FactsAreImportant-Medication AbortionReversal.pdf

7. Gordon, Mara. "Controversial 'Abortion Reversal' Regimen Is Put To The Test" *National Public Radio.* (2019)

   https://www.npr.org/sections/health-shots/2019/03/22/688783130/controversial-abortion-reversal-regimen-is-put-to-the-test?fbclid=IwAR1n613V0tWOTZyENqXYtHNqyiY0aZt BEc_xa0f3qzNIdnRL_InA98DWK9A

8. Rettner, Rachael. "Abortion Pill Gets New Label: 5 Things to Know About Mifepristone" *Live Science.* (2016)

Abortion Pill Reversal

https://www.livescience.com/54238-abortion-pill-mifepristone-label.html

9. Delgado, George. et al. "A Case Series Detailing the Successful Reversal of the Effects of Mifepristone Using Progesterone" *Issues in Law & Medicine.* Volume 33, Number 1 (2018)

https://issuesinlawandmedicine.com/wp-content/uploads/2018/10/Delgado-Revised-09-2018-1.pdf

10. Ibid

11. "SFP research grant awards" *Society of Family Planning.* (2018)

https://www.societyfp.org/Research-and-grants/Grants-funded.aspx

12. Spears, Larry. "Warning Letter" *Food and Drug Administration.* (2002)

http://abortiondocs.org/wp-content/uploads/2014/03/Creinin-Mitchell-FDA-Warning-Letter-6-12-2002.pdf

13. Gordon

14. "SFP research grant awards"

The Procedures

THE FORCEPS

Chapter 5

Forceps Abortions & Aborting Disabled Fetuses

Then I inserted my forceps into the uterus and applied them to the head of the fetus, which was still alive...

Late-Term Abortion Specialist Dr. Warren Hern[1]

When I first saw the type of forceps used for abortion, it was an instrument that overwhelmed and disturbed me more than any other instrument. Most people find the abortion forceps to be the most upsetting. This instrument looks much more overtly violent than the curettes that I had seen previously. Seeing this instrument is a turning point for many people. It's the moment when you realize that someone obviously designed this instrument with death and destruction as its goal.

The Procedures

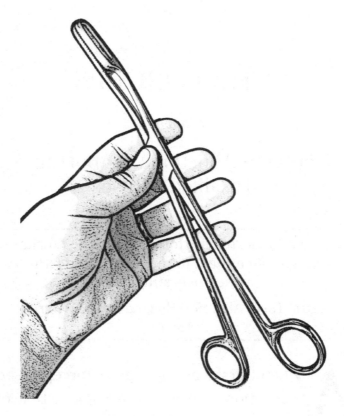

Sopher Ovum Forceps

The forceps used in forceps abortions are not like those used to aid the delivery of babies. Forceps come in many different shapes and sizes depending on the task for which they were designed. The forceps used for delivering babies have large hoops on the end that wrap around the baby's head. Some have described it as looking like large salad tongs. It doesn't look nearly as menacing. These are not the forceps used in abortion. The only time this type of forceps would be used in an abortion would be for an abortion near the end of pregnancy where the dead baby is delivered whole and the forceps are needed to aid delivery.

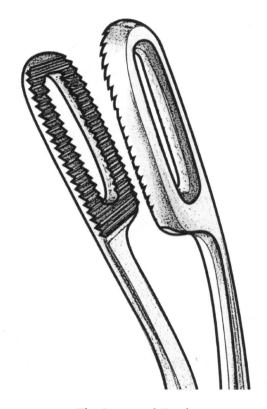

The Jaws and Teeth

The forceps used for abortions instead have smaller hoops or jaws that are lined with sharp teeth. It looks a lot like a needle nose pliers except that the teeth are larger and sharper. The jaws are about the size of a large thumb. Abortion forceps come in many shapes and sizes as well, but they all have several

The Procedures

characteristics in common. They are all made of rigid metal. They all have handles and a hinge in the center allowing the doctor to open and close the jaws. And they all have the smaller hoops lined with sharp teeth. Forceps may have other features as well, such as a ratchet. The ratchet is used to keep hoops closed tight until the doctor is ready to open them. This way an abortionist can grab a body part and it won't let go.

While some forceps used in abortion aren't necessarily designed for abortions, many of them are. One such forceps is the Hern Ovum Evacuation Forceps.[2] This forceps takes its name from Dr. Warren Hern, one of the most well-known and outspoken late-term abortionists in the country. Hern's practice in Colorado performs abortions well into the third trimester. The word "ovum" refers to the egg. And the word evacuation simply refers to removing something from the uterus. I don't know who had the idea to call it an "ovum evacuation" forceps instead of a fetus evacuation forceps. The forceps is never used to remove an egg, fertilized or unfertilized, from the uterus. It isn't even used to remove embryos. It is only used to remove fetuses at the very end of the first trimester and later. The Hern forceps is made specially with the shape and size he prefers. I can't imagine being such a prolific killer as to have a deadly instrument made exactly to my specifications.

The Forceps

The forceps are generally used in one of two ways: to remove pieces or to crush. To remove pieces is relatively straight forward. You simply grab whatever body part you can find, an arm or leg for example, and pull it off. You must continue removing pieces until all that is left is the head. This method of pulling off body parts is typically done in the second trimester. In the first trimester, the body is small enough as to not need a forceps. And in the third trimester, the skeletal system has hardened to the point where it is very difficult if not impossible. However, if the fetus is killed with a lethal injection the day before, decomposition may aid in removing pieces.

Most of these abortions occur in the second trimester when it is too large to be vacuumed out but not too large to remove in pieces. In most second trimester forceps abortions, the fetus dies when it bleeds out as a result of losing limbs. It dies in much the same way that you or I would die if our limbs were removed. In the second trimester, it is hard work to remove pieces. It takes substantial arm strength to get the job done. The sharp teeth help to get a good grip. When all that is left is the head, then the head must be crushed. The head is the greatest challenge as it is large and round. One way to deal with this problem is to crush it with the forceps. Then the head that was

The Procedures

crushed must be scraped out with the sharp uterine curette. The crushing results in sharp skull fragments. These skull fragments must be removed carefully so as not to damage the uterus. Removing sharp skull fragments is risky.

In addition to removing pieces, the forceps are also used for crushing. Not only are they used for crushing the head but also the rest of the body. The sharp teeth of the forceps are especially important when crushing. With this method, the abortionist simply crushes up the body parts till they are nearly unrecognizable and then pulls them out. One of the downsides of crushing is that it doesn't preserve body parts which can then be legally donated or illegally sold for research. The perceived benefit of crushing is that the abortionist and staff don't have to look at the recognizable body parts. It allows the abortionist to dehumanize the child so that he doesn't have to confront the violence so directly. It also requires less strength to remove pieces if they have been crushed up first.

Another downside to crushing with the forceps is that it is harder to confirm that the abortionist removed everything. Normally the clinic staff has to examine the parts with a backlight to make sure they got everything.

The Forceps

If parts are left in the uterus, it can become infected and further damage can be done to the uterus. This is especially true if pieces of the skull are left behind. Skull fragments are sharp. Sharp objects in the uterus can become dangerous very quickly. One technique used to overcome the problem of examining tissue after it has been crushed is weighing the tissue. The abortionist uses measurements from an ultrasound to estimate the total weight. If the parts don't weigh enough, then he suspects that there may still be pieces inside.

Remember, forceps abortions are mostly done in the second trimester. Sometimes they are done into the third trimester, but by then the baby has grown very large and the bones have hardened further. Using the forceps in the third trimester is often too difficult and requires a lethal injection abortion which is described in the next chapter. As you now know, the fetus in the second trimester is a fully developed baby. All the parts are in place, including organ systems, face, hands, and feet. The only difference between a second trimester fetus and a third trimester viable fetus or even a birthed newborn baby is size and maturation. All that needs to happen for a second trimester fetus to be born is simply to get bigger and become more mature.

The Procedures

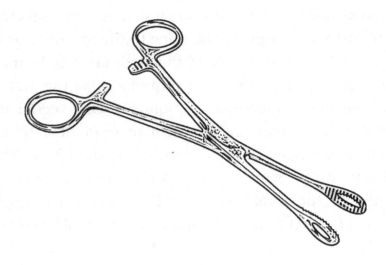

Sponge Forceps

The earliest that a forceps may be used is at the end of the first trimester. When doing a vacuum aspiration abortion, the doctor might find that the head is too big to be vacuumed. He may use a sponge forceps to crush the head and remove it. The sponge forceps is the smaller cousin to the larger forceps used in second trimester abortions. The forceps can be used from the end of the first trimester right up to full term.

Women get abortions for nearly every conceivable reason. Planned Parenthood v. Casey, the Supreme Court decision that replaced Roe v. Wade in 1992, declared abortion a constitutional right for any conceivable reason through 24 weeks gestation. This has been one of the conflicting issues in public opinion.

The Forceps

While half of the country claims to support Roe and its companion cases, most Americans do not agree that you should be able to get an abortion for any reason. Most Americans do think that there should be a justifiable reason. But Roe has made it impossible to even have a public discussion over what justifies an abortion.

This has resulted in two terms that we frequently hear: elective abortions and therapeutic abortions. These terms are widely used in public debate but are also necessitated by medical billing. If your insurance only pays for abortions for certain reasons, you need to have billing codes that allow the doctor to bill correctly for the given situation. For example, Medicare usually pays for abortions only in the cases of rape, incest, or a qualifying medical reason. Elective abortions are generally understood to be those that are done for reasons other than medical reasons. If a woman gets an abortion because she doesn't feel like having a baby, that is an elective abortion. She had a reason for making that decision, but it was not a medical reason. But if a doctor recommends an abortion for a medical reason, that is considered to be a therapeutic abortion. The line between the two, however, is fuzzy at best. What if a doctor recommends abortion because the woman is stressed out about having a baby? Stress could be

The Procedures

considered a mental health condition and have negative health consequences. Something as minor as stress could be used as an excuse to label it a therapeutic abortion. In fact, some abortion advocates go so far as to claim that every abortion is potentially saving the woman's life and therefore every abortion is therapeutic. My position and that of many people in the pro-life movement is that abortion is a horrifically violent act and only justifiable when the mother is given necessary lifesaving treatment that may inadvertently kill the unborn child. The most common example of this is surgery for ectopic pregnancy, as explained in chapter 5. These lifesaving treatments are not done in abortion clinics.

The reasons given for abortion change somewhat once we get into the second trimester and the forceps are used as the primary instrument. This is because the number of fetuses aborted for "fetal abnormalities" increases somewhat later in the second trimester. The term "fetal abnormality" is a euphemism for any fetus with a disability or medical condition. These children are labeled abnormal and often aborted as a result. It is an Orwellian euphemism, as are so many used by the pro-choice movement. We would never say that an adult with Down Syndrome has "adult abnormalities" or that a teen with autism has "teen abnormalities." But any unborn child with a medical condition is labeled abnormal and abortion is offered as the solution. These wide ranging medical conditions and disabilities are often discovered with an ultrasound around 20 weeks gestation. This is one of the reasons why the pro-choice movement fights so hard against 20 week abortion prohibitions. They don't want to lose the ability to abort an "abnormal" baby.

These medical conditions can range from very serious conditions that will inevitably kill the baby to very minor conditions that can be managed or fixed with surgery. Some of these are considered "chromosomal abnormalities" because the chromosomes did not fuse together correctly at conception. Down Syndrome or trisomy 21 is a common example. The 21st chromosome pair has a third chromosome attached causing Down Syndrome. This occurs at conception. Other disabilities are in a category called neural tube defects. These birth defects, which include spina bifida, cleft palate, and club feet occur at six weeks gestation when the embryo does not fold up completely. These are just two categories of disabilities for which people choose to abort. For these babies, instead of being given the dignity of medical care and a natural death, they are violently euthanized with forceps.

The pro-choice movement often jumps to the hard cases, those unborn babies with a terminal condition,

The Procedures

as if a violent forceps abortion is better than a natural death. But many of these babies aborted because of disabilities are not terminal. Whether it's spina bifida or Down syndrome, with adequate medical care and therapy, these children can go on to live very happy and fulfilling lives, so long as they aren't aborted first. Two thirds of Down syndrome babies are aborted in the United States. Of those diagnosed with Down syndrome in utero, 92% are aborted. And yet over 95% of teens and adults with Down syndrome reported that they liked how they look, liked who they are, and were happy with their lives.[3] Almost all of these aborted Down syndrome babies would have grown up to be happy, fulfilled, and differently-abled adults.

Unfortunately, the tests that are used to diagnose babies with Down syndrome in utero put the child at risk of miscarriage or still birth. Parents are not generally encouraged to get the test unless they are willing to abort. That is why 92% of those diagnosed with Down syndrome are aborted. Those who would not abort are not likely to put their baby at risk of miscarriage or still birth with the test. The tragic result is that many healthy babies that do not have Down syndrome die from the test each year. The most common diagnostic test is called amniocentesis with approximately 200,000 tests done each year in the

United States.[4] The parents typically first do a blood test and ultrasound to determine if their baby is at a higher risk of Down syndrome. Amniocentesis is usually only done when there is a higher risk and when the parents are considering abortion. With the risk of miscarriage at 1 in 200 to 1 in 400, that means that approximately 500-1,000 babies are miscarried due to amniocentesis each year in the United States.[5] One published case study found that 93% of the babies that underwent amniocentesis at one clinic were found to have normal chromosomes.[6] If these numbers hold true, as many as approximately 900 chromosomally normal babies are lost to miscarriage and still birth in the United States each year so that their parents can keep open the option of abortion. This shows the wide ranging affects that abortion culture has had on our society.

Even with the increase of abortion of disabled babies in the second trimester, most of these abortions are still purely elective according to late-term abortionist Martin Haskell.[8] In this chapter we've looked at the increasingly violent abortions done generally in the second trimester with the forceps. In the next chapter we will look at lethal injection abortions.

The Procedures

Chapter Notes

1. Hern, Warren. "Did I Violate the Partial-Birth Abortion Ban?" *Slate*. (2003)

   https://slate.com/technology/2003/10/did-i-violate-the-partial-birth-abortion-ban.html

2. "MedGyn Hern Ovum Evacuation Forceps" *MedGyn Products, Inc.*

   https://www.medgyn.com/product/hern-ovum-forceps/

3. Coolidge, Ardee. "The Reason Why So Many People Want to Eradicate Unborn Children with Down syndrome" *Care Net.* (2018)

   https://www.care-net.org/abundant-life-blog/the-reason-why-so-many-people-want-to-eradicate-unborn-children-with-down-syndrome

4. "Amniocentesis" *American Pregnancy Association.*

   https://americanpregnancy.org/prenatal-testing/amniocentesis/

5. Ibid

6. Daniilidis, A. et al. "A four-year retrospective study of amniocentesis: one centre experience." Hippokratia vol. 12,2 (2008): 113-5.

   https://www.ncbi.nlm.nih.gov/pmc/articles/PMC2464303/

7. Brown, David. "Late Term Abortions" *Washington Post.* (1996)

   https://www.washingtonpost.com/archive/lifestyle/wellness/1996/09/17/late-term-abortions/f15ae3a6-9711-45cc-9c13-e5160d293489/?noredirect=on

THE LETHAL INJECTION

Chapter 6

Lethal Injection, Partial Birth, Saline Infusion, and Third Trimester Abortions

---

To stop the heart, potassium chloride is administered directly after the vecuronium bromide. Without proper sedation, this stage would be extremely painful. The feeling has been likened to 'liquid fire' entering veins and snaking towards the heart.[1]

BBC Journalist Ben Bryant

This quote comes from a 2018 article about the execution of criminals in the United States. In this article, he describes a three drug regimen that is used for executions and ends with a fatal dose of potassium chloride. Potassium chloride stops the heart. In essence, the criminal dies of a heart attack, if he hasn't already been killed by one of the other drugs. Potassium chloride doesn't just stop the heart. It also

The Procedures

causes violent muscle spasms, sending the criminal's body into convulsions.[2] This is why the criminal is given a drug to paralyze him.

Not every criminal prefers to be executed with potassium chloride. Some criminals are now asking to be executed with Digoxin. Digoxin is a heart medication used for heart failure and irregular heartbeats. But it is also being used off-label in physician-assisted suicide due to its affordability.[3] Off-label simply means that the drug is being prescribed for a purpose other than purpose for which the FDA approved the drug. Here again, an overdose of Digoxin causes the heart to stop. Not only is potassium chloride used in a cocktail of drugs, digoxin is also used as part of a cocktail in assisted suicides. The reason some criminals prefer to be executed with Digoxin is because it can be taken orally instead of with a lethal injection.[4]

Whether using potassium chloride to execute criminals or digoxin to commit suicide, additional drugs are used to try to make the death less violent and less painful. Drugs are used to render the person unconscious so that they don't suffer as much. And they are also used to paralyze people so that they don't thrash about violently.

Lethal drugs aren't just used on criminals and people who qualify legally for physician-assisted suicide. These lethal injections are also used in the

The Lethal Injection

United States and around the world for third trimester abortions. Occasionally they are also used for second trimester abortions as well. These third trimester abortions are often called lethal injection abortions or heart attack abortions by those of us who are pro-life. It is hard to comprehend that people get third trimester abortions. Many people don't even understand that it is legal in some states to get third trimester abortions. Second and third trimester abortions are often referred to as "late-term abortions." Late-term abortion is a term widely used and understood among English speakers. It is not a technical medical term. Instead, it is plain language. The pro-choice movement hates this term because late-term abortions are overwhelmingly unpopular with the public. Instead they have come up with their own term, "later abortion." I'm not sure why the pro-choice movement thinks later abortion will be a more favourable term for their side. It seems likely to me that the pro-choice movement will have to dump this term as well in favour of something more ambiguous.

The reason it has become common to use the lethal injection in third trimester abortions is because these babies have a chance of survival outside the womb. In the third trimester they are past the generally considered point of viability at 24 weeks gestation. As we learned in chapter 10, their lungs have developed enough that they may be capable of breathing on their own. In the event that a woman wants an abortion of a viable or potentially viable baby, the abortionist wants to ensure that he is successful in killing it. Abortionists do not want to run the risk of an accidental live birth. A lethal injection while the child is still in the womb is the most efficient method devised by abortionists to ensure death. If the same doctor were to give the same lethal injection after birth, it would be considered murder. They have to get the job done before birth. Abortionists typically use digoxin because of its affordability, but some also use potassium chloride. Unlike criminals, however, no other drugs are used to make the death of the child less violent. No drugs are used to render the child unconscious or paralyzed. Even many on the pro-choice side will admit that the child is conscious and able to feel pain after 24-27 weeks gestation.

Pro-choice advocates will often claim that these abortions are rare as justification for third trimester abortions. They point to the fact that these abortions make up just over 1% of all abortions.[5] In my opinion, the fact that these abortions exist at all is unconscionable. I don't consider "rare" to be a justification. Instead, "rare" is an admission that these abortions do occur. So how many of these third trimester abortions are being performed? The best data available is from the CDC's

The Lethal Injection

last abortion surveillance report from 2015. They reported a total of 5,597 abortions at or after 21 weeks gestation in 2015, making up 1.3% of total abortions. Twenty one weeks of gestation isn't considered viable but is the point when some micro preemies have a chance of survival after birth. Even at this "rare" percentage, the number is staggering. Delaware reported seven of these abortions for .3% of all Delaware abortions. The second highest state that reported was Colorado with 298 third trimester abortions at 3% of total abortions. Colorado was higher because that is where Warren Hern specializes in late-term abortions. The highest state was New Mexico with 336 third trimester abortions at 7% of total abortions. New Mexico also has a facility that specializes in late-term abortions.[6]

The real number of third trimester abortions, however, is far higher than the 5,597 reported. Eleven states and the District of Columbia did not report their numbers. States that do not report tend to be much more permissive of late-term abortions. These include California, New York, and Maryland, which are home to late-term abortion facilities. This means that the real number of late-term abortions are sure to be much higher than reported to the CDC. The percentage of total abortions is surely higher than 1.3%. And the total number of these abortions is likely between 10,000-

The Procedures

15,000. These numbers are likely to rise as a number of liberal states have expanded access to third trimester abortions after the election of President Trump.

The state of New York shocked the nation when they legalized third trimester abortions and Governor Cuomo ordered the One World Trade Center to be lit in pink to celebrate. Delaware was the first state to expand late-term abortions in the wake of the election of President Trump when it passed Senate Bill 5 in 2017. It's an unfortunate distinction to be labelled not only "the first state" but also the first state in late-term abortions! Delaware didn't get much notice because we are a small state but also because the governor signed Senate Bill 5 behind closed doors and without any celebration. New York, on the other hand, celebrated loudly with cheering at the signing ceremony and by lighting the One World Trade Center in pink. The celebration and brazen support of third trimester abortions was the wakeup call that the pro-life movement needed.

The pro-choice movement also justifies these late-term abortions by claiming that nearly all of them are to protect the mother's health or because the fetus has a severe or fatal birth defect or medical condition. They never cite statistics or proof that this is why people have late-term abortions because no such proof exists. The best they can do is to give some examples of

The Lethal Injection

women who have had abortions for those reasons. What little data we have suggests that this is not true. Florida, for example, tracks the reasons that people claim they are having abortions. They track by trimester. There were only two abortions in the third trimester reported in 2018. That is not enough to make any kind of statistical claims. But 4,268 second trimester abortions were reported. Only 10.4% were because of a fetal medical condition or disability. And only .7% were because the mother's life was in danger.[7]

A study published in 2013 in the journal *Perspectives on Sexual and Reproductive Health* looked at the reasons women wait till after 20 weeks to get an abortion. Fetal medical conditions and the health of the mother were so insignificant as to not even show up in their research. The authors wrote, "But data suggest that most women seeking later terminations are not doing so for reasons of fetal anomaly or life endangerment."[8]

Further, we can look at those doctors who specialize in late-term abortion to see what they say are the reasons women get late-term abortions. Occasionally, they are honest enough to admit that many of these late-term abortions are elective abortions. Martin Haskell is one of the few doctors to do abortions in the third trimester. In the early 90s, he was quoted in a medical publication as saying, "I'll be quite frank: most of my abortions are elective in that

20-24 week range. In my particular case, probably 20 percent are for genetic reasons. And the other 80 percent are purely elective."[9]

Lethal injection abortions are a much more lengthy and costly process than most other abortions. The woman first has to find an abortion clinic that does third trimester abortions. There are only a handful of clinics around the country that specialize in these risky procedures. She has to go to the clinic for an evaluation and consultation. The abortionist has to examine her, which includes an ultrasound to attempt to determine how large and how old the fetus is. The abortionist then has to discuss which options for the abortion he is willing to do. There are different variations of lethal injection abortions. Once it is determined what variation of abortion they are doing, she has to sign the consent forms consenting to the abortion. She also has to have the money to pay for it. Third trimester abortions cost thousands of dollars. If she has the money, they can start the abortion procedure.

The abortion begins with the lethal injection. The abortionist will have a large needle to inject the fetus with his drug of choice. This abortion requires that the needle be guided by an ultrasound. The needle can either be injected through the abdomen or transvaginally. If it is done through the abdomen, the abortionist simply pushes the needle through the

The Lethal Injection

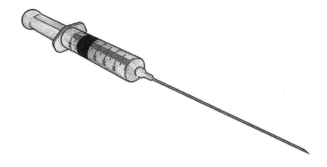

Needle Used for Lethal Injection Abortions

woman's belly till he reaches the fetus to inject it with the drug of choice. Transvaginally means that the abortionist inserts the needle through the vagina and the cervix to reach the fetus. The goal is to get the needle into the body of the fetus to inject the fatal drug. Even if the needle misses and it is injected into the amniotic fluid, it can still successfully kill the child. This point in the abortion can be a particularly traumatic time as the abortionist and the mother are likely to feel the baby struggle as it is injected. At this visit the abortionist will also insert laminaria to begin dilating the cervix. Laminaria is a seaweed that absorbs fluid and expands to dilate the cervix.

This second appointment typically occurs the next day. This appointment begins with an ultrasound to confirm that the fetus is actually dead. Sometimes if the abortionist fails to inject the fetus, it may survive. Typically if the abortionist misses the fetus and injects

The Procedures

the drug into the amniotic fluid, it will still kill the fetus. But it will also take significantly longer to kill it. And occasionally, the doctor discovers at the second appointment that the fetus survived. In this case, the lethal injection must be administered a second time. You will remember from chapter 4 that Dr. Kermit Gosnell was doing these lethal injection abortions, but was not confirming that they were dead. Instead, if the baby happened to be alive after it was born, he would kill it after birth. You will also remember from chapter 4 that these abortions were done across state lines in order to avoid being charged with illegal late-term abortions.

Assuming the fetus was successfully killed, the next step is to remove it. This is done typically in one of two ways: through induced labor or through dismemberment. Induced labor simply means that labor is induced and the woman births an intact and dead baby. Dismemberment means that the abortion is finished with a forceps. The abortionist removes the fetus piece by piece, the same as a second trimester forceps abortion. Some abortionists believe that the bones soften by the next day after the fetus died, making it easier to dismember.[10] Sometimes the abortionist may give the mother the option of which method to do. Forceps abortions in the third trimester are at high risk for uterine perforation due to the large size of the fetus and the size of the uterus which has been stretched to accommodate the child.

Lethal injection abortions are often confused with partial birth abortions, which are done in the second and third trimester. I frequently get questions during my Fetal Beauty presentations about partial birth abortions. Lethal injection abortions are not the same as partial birth abortions. Partial birth abortions are a crime under federal law and also under many state laws. A partial birth abortion is when a living fetus is delivered all but the head. A suction tube is inserted into the back of the head. The brain is removed causing the head to collapse. The primary reason partial birth abortions were done was to solve the problem of the head. The head is a challenge to abortionists because it is large and round. By collapsing the head, it makes the fetus easier to deliver and doesn't result in sharp skull fragments inside the uterus. The textbook *Management of Unintended and Abnormal Pregnancy* describes the head problem stating, "Because the cranium represents the largest and least compressible structure, it often requires decompression."[11]

There are many other variations of the procedure that fall within the federal legal definition of partial birth abortion, one of which causes the death of the fetus by decapitation. A scissors or forceps is used to

The Procedures

separate the head from the body while the head is still in the birth canal.[12] Congress made partial birth abortion a crime in 2003 and the law was upheld by the Supreme Court in 2007. However, an abortion is not considered a partial birth abortion if the fetus is already dead before entering the birth canal. This is where lethal injection abortions come in. A lethal injection can cause the death of the fetus before partially delivering the child. There are other tricks that abortionists use as well. An abortionist may remove the arms and legs before doing the partial birth abortion. By removing limbs, it is impossible to prove whether or not the fetus was alive when it was partially delivered. Another method is to cut the umbilical cord first to cause the death of the fetus before partially delivering it. The generally accepted alternative, however, has been lethal injection abortions. According to the Society of Family Planning, lethal injection abortions are done widely "to avoid signs of life at delivery."[13]

Another late-term abortion procedure often confused with lethal injection and partial birth abortion is the saline abortion. This type of abortion is also called an infusion or instillation abortion and can use saline or other toxic substances to kill the fetus. This abortion procedure is rare in the United Sates today but was much more common in the 1970s. Many of the survivors of failed abortions that we hear about today

The Lethal Injection

were survivors of saline abortions. In this abortion, some of the amniotic fluid is removed to make room for the toxic substance. The saline solution or other toxin is injected into the amniotic fluid. A healthy fetus practices breathing in the womb. Instead of breathing air, she breathes amniotic fluid. But in a saline abortion, she ingests the toxin. Saline solution has a burning effect both on the inside and outside of the fetus. The result is a horribly disfigured body. After the death of the fetus, the woman goes into labor and delivers the dead fetus. Sometimes lethal injection abortions accidently become infusion abortions. If the doctor misses the body of the fetus with the injection, he may then accidentally inject the drug into the amniotic fluid. The drug still kills the child, but it is killed through infusion into the fluid rather than injection into the body. As I'm sure you can see at this point, there are many variations in the ways that doctors do abortions. Abortions don't fit neatly into categories. What they all have in common is that they are all violent acts.

Third trimester abortions are the most violent abortions and also the abortions the public overwhelmingly wants to prohibit legally. This is true due to the fact that the larger and more developed the fetus is, the more violent the abortion must be to successfully kill and remove it. We know that all abortions are violent in so far as all are acts of killing a

The Procedures

living human being. But the amount of bodily injuries done increases as the child grows. By the time we get into the third trimester, it is hard to judge which abortions are more violent than others. For example, is an abortion that involves crushing of the head with a forceps more or less violent than one that involves aspirating the brain matter? I would say that crushing with the forceps is more violent as it destroys the outside of the head and skull more than just aspirating the brain. But that's just my subjective opinion. I think what most people can agree with is that the lethal injection is less violent than other forms of second and third trimester abortions. A drug used to stop the heart seems to be a less violent way to kill the fetus because the amount of injuries done to the body are less than dismemberment with a forceps, aspiration of the brain, crushing of the head, or decapitation. This is the motivation behind several recent state laws that prohibit the killing of the fetus through dismemberment. The goal of these laws is to force abortionists to use lethal injection to kill the fetus in the same way that the partial birth abortion ban has forced abortionist to resort to the lethal injection. The motivation is to force abortion doctors to use a less violent method of killing the fetus.

While the lethal injection seems less violent than other late-term abortions, it is still a more violent procedure than the lethal injection used on criminals. No drugs are used to render the fetus unconscious or paralysed. As a result, the fetus is likely to thrash about violently. The amount of thrashing is likely to increase further into pregnancy as the fetus grows larger and stronger. Also, the lethal injection doesn't stop further violence after death. While injuries to the body after death wouldn't be considered violence in the same way as injuries before death, those injuries are still significant. It is still a human body that is being destroyed. In our culture, the human body, even though death has occurred, it still worthy of a certain amount of dignity and respect. This is why states have laws against the desecration of bodies. This is also why the State of Indiana passed a law requiring the burial or cremation of fetal bodies after abortion. That law was upheld by the U.S. Supreme Court in 2019 in Box v. Planned Parenthood of Indiana and Kentucky Inc.

Another aspect of late-term abortions that is widely misunderstood is the fact that sometimes babies are accidently born alive after botched late-term abortions. Also misunderstood is what is done with those babies after they are born. Many on the pro-choice side refuse to admit that this occurs. They don't want to admit that babies can survive failed abortions, especially viable babies. But we know that this occurs because these babies have grown up to be adults and those adults are speaking out. A few of the more well-known survivors of botched abortions include Gianna Jessen, Melissa Ohden, Claire Culwell, and Josiah Presley. I would encourage you to learn more about each of these survivors. Each survivor has an incredible story of how they beat the odds and are alive today to talk about it.

There are a large number of ways that people can survive abortion attempts. They can be loosely categorized into two groups: those that survived failed abortions before viability and went on to develop to full term and those that were born alive during a third trimester abortion gone wrong. The first group can include situations where the fetus survives the abortion pill, where the abortionist fails to kill and remove the fetus, or where the abortionist fails to detect the presence of another fetus when there are twins or multiples. When an abortion fails, the pregnant mother may decide not to go back for a second abortion to finish the job. One of the more famous examples is that of Claire Culwell who survived when her mother went for a surgical abortion. The abortionist got her twin but missed her.[14] You can visit www.ClaireCulwell.com to learn more about Claire's story.

But for this chapter we are more concerned about babies that are accidently born alive during a third trimester procedure or at the end of the second

The Lethal Injection

trimester. These are babies that may have a chance of survival if given the best level of care in a NICU. There are three things that can happen to a baby that is born alive: the doctor can send it to a NICU and attempt to save it, the doctor can refuse care until it dies, or the doctor can actively kill it. In 2002, Congress passed and President George W. Bush signed into law the Born Alive Infants Protection Act which declared that babies born alive after a failed abortion should be treated equally under the law.[15] Under this law, doctors are obligated to treat these babies the same as any other premature baby. But under the Born Alive Infants Protection Act, there are no teeth to punish abortion doctors who refuse to give these babies born alive after a failed abortion an equal chance at life. In 2019, Republicans in Congress made numerous attempts to add teeth to this law, but failed as pro-choice Democrats opposed these efforts to strengthen the Born Alive Infants Protection Act. I'm not pointing this out for partisan reasons as I am myself an active member of the Democrat party. I do hope that my own party will come to understand that true progress and true equality will move us away from abortion violence.

Without any enforcement of the Born Alive Infants Protection Act, if a baby isn't given NICU care, it may simply be refused care until it dies. Unfortunately, this seems to be the method of choice for many abortionists.

The Procedures

Even if he violates the Born Alive Infants Protection Act, he won't be prosecuted or punished. Babies at 24 weeks gestation have a greater than 50% chance of survival with proper NICU care. But without a NICU, they cannot survive on their own until several weeks later. And so babies that are viable and very likely to survive can be killed by simply denying care. Many individuals have given their testimonies about witnessing babies that were denied care. One such woman is a nurse who I know personally. Leslie Dean wrote:

> *The doctor estimated him to be between 19-20 weeks. His body had been badly burned, and the expression on his face was unmistakably one of intense pain. He was still alive.*
>
> *The doctor explained if the eyes were not "fixed" we may need to resuscitate. As he held up the baby to check the eyes, the mother saw him and began to scream uncontrollably: "Oh, God, what have I done?"*
>
> *Declaring the eyes were fixed, he dropped the baby in a bucket on the floor where I saw it moving and gasping for breath, and then died.*[16]

The Lethal Injection

The third option is to actively kill the child outright. The late-term abortionist Kermit Gosnell is spending life in prison for actively killing babies after they were born alive. Actively killing after birth is not as common due to the risk of being prosecuted and jailed for murder. But there have been testimonies by witnesses of other doctors doing similar killings after babies were born alive. A doctor might decide to actively kill the child if it could survive without the care of a NICU or if it is necessary to harvest fresh organs.

The pro-choice movement is currently very opposed to the idea of making abortion doctors save babies that are born alive. They now frequently claim that babies aren't born alive, that late-term abortions are not real, and that babies aren't being denied care. At the time of this writing, these types of claims have been made by a number of the Democrat candidates for President. Unfortunately, there are very few statistics about how many babies are born alive after failed abortions and what happens to those babies. There are some reports out of Canada, Australia, and the United Kingdom. But those reports aren't thorough enough and reliable enough to give us a good picture of how many of these babies exist. We do know that babies are born alive after failed abortions because of those few reports and because of the stories told by witnesses. But we also know that babies are born alive after failed abortions because the abortion industry on rare instances admits to it. In *Management of Unintended and Abnormal Pregnancy* the authors state that one of the reasons abortionists might use the lethal injection is because "they desire to avoid the possibility of unscheduled delivery of a live fetus."[17]

This has been one of the longest and most difficult chapters to write. I can't tell you how much I appreciate you putting yourself through the emotional pain of reading this chapter. There were many mornings when I wrote only a few sentences before I had to put it down and walk away. But it's incredibly important that everyone understand what our society has allowed to happen since Roe v. Wade and Doe v. Bolton. And it is equally important that every one of us stand up to the pro-choice movement and put an end to this. This is happening on our watch.

The Lethal Injection

Chapter Notes

1. Bryant, Ben. "Life and Death Row: How the lethal injection kills" BBC. (2018)

   https://www.bbc.co.uk/bbcthree/article/cd49a818-5645-4a94-832e-d22860804779

2. Ibid

3. Aleccia, JoNel. "Northwest doctors rethink aid-in-dying drugs to avoid prolonged deaths" *The Seattle Times*. (2017)

   https://www.seattletimes.com/seattle-news/health/northwest-doctors-rethink-aid-in-dying-drugs-to-avoid-prolonged-deaths/

4. Harcourt, Bernard. "Second Amended Complaint." Hamm v. Dunn. Filed in the United States District Court for the Northern District of Alabama. (2018)

   http://blogs.law.columbia.edu/update-hamm-v-alabama/files/2018/03/94-1-Hamm-Second-   Amended-Complaint-STAMPED.pdf

5. "Later Abortions" *Guttmacher Institute*. (2017)

   https://www.guttmacher.org/evidence-you-can-use/later-abortion

6. Jatlaoui, T. et al. "Abortion Surveillance – United States, 2015" *Centers for Disease Control and Prevention*. (2018)

   http://dx.doi.org/10.15585/mmwr.ss6713a1

7. "Reported Induced Terminations of Pregnancy (ITOP) by Reason, by Trimester" *Agency for Health Care Administration*. (2019)

   https://ahca.myflorida.com/MCHQ/Central_Services/Training_Support/docs/TrimesterBy Reason_2018.pdf

8. Foster, Diana and Katrina Kimport. "Who Seeks Abortions at or After 20 Weeks?" *Wiley Online Library*. (2013)

   https://doi.org/10.1363/4521013

The Procedures

9. Brown, David. "Late Term Abortions" Washington Post. (1996)

   https://www.washingtonpost.com/archive/lifestyle/wellness/1996/09/17/late-term-abortions/f15ae3a6-9711-45cc-9c13-e5160d293489/?noredirect=on&utm_term=.4fba3ce7b6b1

10. Paul, Maureen, et al. *Management of Unintended and Abnormal Pregnancy* Hoboken: Wiley-Blackwell, 2009, pp. 166

11. Paul, pp. 173

12. Gorney, Cynthia. "Gambling with abortion: why both sides think they have everything to lose." The Free Library. (2004)

    https://www.thefreelibrary.com/Gambling+with+abortion%3A+why+both+sides+think+they+have+everything+to...-a0126194929

    See exchange between DOJ lawyer Quinlivan and Physician "Doe." This exchange describes decapitation with a scissors. I've also seen video of decapitation with a forceps.

13. Diedrich, Justin and Eleanor Drey. "Clinical Guidelines: Induction of fetal demise before abortion" *Society of Family Planning.* (2010)

    https://www.societyfp.org/_documents/resources/Induction ofFetalDemise.pdf

14. Culwell, Claire. "My Story"

    http://www.claireculwell.com/my-story.html

15. H.R. 2175. "An act to protect infants who are born alive." 107th Congress, 2D Session. (2002)

    http://www.nrlc.org/uploads/bornaliveinfants/Baipatext.pdf

16. Dean, Leslie. "I've Had 2 Abortions. Here's Why I Support Alabama's Pro-Life Law." *The Daily Signal.* (2019)

    https://www.dailysignal.com/2019/05/19/ive-had-2-abortions-heres-why-i-support-alabamas-pro-life-law/

17. Paul, pp. 169

DARE TO DREAM

Chapter 7

Why We Should End Abortion Violence

---

*There is a balm in Gilead
To make the wounded whole;
There is a balm in Gilead
To heal the sin-sick soul.*

*Sometimes I feel discouraged,
And think my work's in vain,
But then the Holy Spirit
Revives my soul again.*

*If you cannot sing like angels,
If you can't preach like Paul,
You can tell the love of Jesus,
And say He died for all.*

*Traditional African-American Spiritual
Author Unknown*

Ending abortion in America isn't as big of a dream as it first appears. Unfortunately, most people seem to have accepted abortion as just a fact of life. But that's not supported by the data or our history. In fact, legalized elective abortions are only a recent phenomenon in American history. America is about

The Procedures

243 years old as of the writing of this book. But Roe was decided by the Supreme Court only 46 years ago. Further, the abortion rate is at a historic low and continuing to fall. The abortion industry is in crisis as fewer and fewer people choose abortion.

In this chapter we are going to look at recent societal trends and how our culture is moving away from abortion. Then we are going to put abortion in the context of our history of oppression of African-Americans and our nation's core values.

In order to understand how we got here and where we are going, we need to look at the big picture starting with the birth control pill and the sexual revolution. Abortion didn't just come out of nowhere. There were events that led to it. Before the introduction of the birth control pill in 1960, the only methods of birth control were fertility awareness and barriers, especially condoms. The problem with condoms is their high failure rate. About 18 out of every 100 women using condoms will get pregnant in a year. And so casual sex and premarital sex came with a high cost. The risk of getting pregnant and the social and economic costs of pregnancy outside of marriage meant that people were less willing to engage in sex outside of marriage. But then the birth control pill came along with a failure rate half that of condoms. Almost overnight the sexual

98

revolution was born. The birth control pill was just one factor, but it was a very important factor in that people perceived lower risk.

But in reality, the birth control pill had a substantial failure rate. The result was a lot more people having sex with the birth control pill which has a failure rate of 9 out of every 100 women in a year. Not surprising the result was a wave of unwanted pregnancies and along with those unwanted pregnancies a demand for abortion. Women wanted a backup plan in case birth control didn't work. That back up plan was abortion. By the time Roe was decided in 1973, the abortion rate hit 16.3 abortions per 100,000 women. Roughly half of those were illegal and half were legal. A few states had legalized elective abortions before Roe. But it opened the floodgates. In the decade following Roe, the abortion rate nearly doubled. Notice that legalizing abortion caused the abortion rate to go up dramatically, not down. It's really shocking and outrageous that the abortion industry has convinced millions of Americans that legalizing abortion causes the abortion rate to go down. That's simply a lie. The abortion rate did not go down after Roe. It doubled. The abortion rate only started to decline around 1990. The abortion rate continued to rise for 13 years following Roe.

The Procedures

In 1990 that all changed. Not only did the abortion rate begin to fall, it fell dramatically. And it continues to fall to this very day. 2014 marked a very important year since that was the year that the abortion rate fell below that of 1973 when Roe was decided. In other words, for the first time since Roe, the demand for abortion that resulted in Roe just isn't there anymore. This terrifies the abortion industry. Their revenue is at stake. Planned Parenthood has been able to keep their revenue up by taking a greater market share of abortions away from the independent clinics. But more importantly, if increased demand for abortion resulted in Roe, why couldn't a decreased demand result in the overturning of Roe? I think that the decreased demand certainly could lead to the end of Roe. And I think many in the abortion industry would quietly agree with me, even if they don't want to admit it. The reality is that we live in a Democratic Republic. All three branches of our federal government are reactive to the culture. Politicians can pass whatever laws they want now, but the culture always ultimately wins. Politics, entertainment, and media are important parts of the culture. But they are not as important as peoples' everyday decisions about how to live their lives. There is no greater change in the culture regarding abortion than the precipitous and decades-long decline in abortion. It hasn't hit politics yet, but people are voting with their pregnancy decisions. They are voting against abortion.

Dare To Dream

It is beyond the scope of this book to look comprehensively at the reasons for the abortion decline. And in some ways it really doesn't matter. The fact is that abortion is going out of style. The culture isn't changing back to that of the 1950s, but it isn't staying stuck in 1973 either. It is changing into a new 21st century culture. That culture is clearly rejecting abortion. Among the many factors and proposed factors in this decline include contraceptives, social acceptance of single parenting, the economy and prosperity, pornography, delayed sexual activity in young people, TV shows depicting teen parents, and better ultrasound technology. There is even a hypothesis, inspired by the Donohue – Levitt hypothesis linking abortion to crime, which says that the abortion rate has gone down as a result of what would have been future abortive women being aborted themselves. While I don't know if this hypothesis has much evidence, it would be shocking if the abortion industry was actually aborting away its future customers and putting itself out of business. The pro-choice movement puts a singular emphasis on contraceptives and sex education. They often cite the Affordable Care Act and newly developed hormonal contraceptives as reason for the decline. While I'm sure that contraceptives are a factor, it seems far-fetched to give so much credit to contraceptives when abortion has been in a steady

The Procedures

consistent decline since 1990 and our culture has been flush with contraceptives going back to the sexual revolution. The pro-choice movement usually fails to give any consideration to all the other factors while maintaining a singular focus on contraceptives.

There is one glaring factor in the abortion decline, however, that is not tied to contraceptives: the number of young people increasingly delaying sex. Not only is abortion going out of style, but so is sex among young people. Kate Julian wrote a fantastic in-depth article in *The Atlantic* entitled "Why Are Young People Having So Little Sex." She begins, "Despite the easing of taboos and the rise of hookup apps, Americans are in the midst of a sex recession."[1] This is not a conservative or pro-life article. It is published in the progressive magazine *The Atlantic*. If you want to further understand the sexual recession, I'd encourage you to start with this article. There is no doubt that we are in a dramatic sexual recession, one that mirrors the decline in abortion beginning in 1990. Today it is more likely than not that an average high schooler is not having sex. That's really shocking. In fact, the decline in teen sex and in abortion may be one of the greatest recent social advancements in our country, despite the fact that most people are unware of it. In other words, we went from a shocking sexual revolution to an equally shocking sexual recession. But this doesn't fit in to the

Dare To Dream

contraceptive narrative coming from the pro-choice movement and their friends in the media.

The reasons for both of these declines are largely unknown. This falls into an area of social science that nobody seems to be able to pin down. Like many people, I have my own hypothesis which I will share with you. It is a relatively simple idea. I believe that young people are rejecting their parents' life decisions. I think young people are smart enough to see their parents' and their grandparents' dumb mistakes. They can see how much they've suffered from those choices and they simply don't want to make the same mistakes. And so they are going to the other extreme. Not only are they putting off sex and rejecting abortion, they are also putting off marriage or rejecting marriage entirely. Am I giving young people too much credit? Maybe I am, but I don't believe so. Are young people overreacting? Maybe in some ways they are. I love being married. I wish everyone could have a marriage and parenting experience that is as fulfilling as mine. But it seems that many young people have checked out of that whole idea.

Regardless of how you see it, the future is very bright for the pro-life movement. We have never been in a better position to make gains across all areas. We will win because the culture is changing. It's not

The Procedures

changing back to the predominantly Christian culture of the 1950s or the abortion culture of the 1970s. It is a new 21st century culture. It is a different culture. It is unbelievably prosperous. It is more supportive of single parents than ever before.

Ultimately, I believe we will win because Americans believe in equality and reject violence. What you will see in the last section of this book is that America was founded on certain core values and that these values cannot be consistent with abortion violence.

On Saturday September 26, 2015, Felicia and I took a day trip to relax on Jekyll Island just off the coast of Brunswick, Georgia. On Monday, the baby we were adopting was to be delivered via a scheduled cesarean section. But on that Saturday we were trying to calm our nerves and mentally prepare. Jekyll Island is a popular vacation destination today. But it also has a long history of slavery and racial segregation. And at one time it was the location of some of the last gasps of the trans-Atlantic slave trade.

In the mid-19th century, Jekyll Island was owned by the DuBignon family. They were slaveholders and proponents of slavery. They also conspired to illegally smuggle slaves from Africa. This was the time period right before the American Civil War. Slavery was still legal in the southern states, including Jekyll Island. The trans-Atlantic slave trade, however, was illegal in America, including the south. This didn't stop proponents of slavery from trying. They aggressively believed it was their right to purchase slaves in Africa and bring them to America. One such proponent was Charles Lamar. Lamar conspired with the DuBignons and a man named William Corrie, the owner of a pleasure yacht named *Wanderer*. Together this group of slavery proponents plotted and successfully carried out a plan to buy slaves at the mouth of the Congo River. The ship was retrofitted with a secret deck for the slaves and a secret fresh water tank. But they needed a private place to unload the slaves away from prying eyes. Jekyll Island provided just the place. *Wanderer* was the last illegal slave ship to deliver to Georgia and one of the last slave ships to deliver to America. The federal government tried and failed to prosecute the group. Not long after, the nation would enter into the most deadly war in American history. Ironically, the *Wanderer*, which was used to enslave over 400 people, was converted by the Union Army into a war ship to free those same slaves.[2]

For my family, this story is now part of our story. On the Monday after our visit to Jekyll island, our son Derrick was born just a few miles away. Unfortunately,

The Procedures

we don't know his family tree. We don't know the true story of his ancestors and how they came here. But what we do know is that nearly all of his DNA comes from that region of Africa. We also know that in the county in which Derrick was born, 72% of the population were slaves in 1860, immediately before the civil war.[3] And so the history of slavery and segregation has become a part of my family's history through adoption. But America's story doesn't end there. In the span of about 240 years, we went from being a country of slavery to a country of segregation and Jim Crow oppression to a country where my wife and I were able to adopt a child of another race and be accepted as a mixed race family. America was transformed into a country where we are accepted and supported by our church, neighbors, and community.

During the period before the civil war, American abolitionists could be divided roughly into two groups. They were the followers of William Lloyd Garrison who believed that America was founded on slavery and those like Frederick Douglas who believed later in his life that America was founded on principles that were inherently inconsistent with slavery. For Garrison, our Constitution was a pro-slavery document, America was founded as an evil country, and the American government was beyond redemption. They swore off civic activities such as voting and politics. But the

Dare To Dream

former slave and popular public speaker Frederick Douglas came to disagree with Garrison. Instead, he believed that America's founding was good and right. He defended the core values that America was built upon. These core values were best summed up by the founding father George Mason in the Virginia Declaration of Rights which inspired the Declaration of Independence. He wrote, "That all men are by nature equally free and independent and have certain inherent rights, of which, when they enter into a state of society, they cannot, by any compact, deprive or divest their posterity; namely, the enjoyment of life and liberty, with the means of acquiring and possessing property, and pursuing and obtaining happiness and safety."[4] And so we can see that included in our core values are life, liberty, and equality.

Douglass used his soaring rhetoric to call Americans back to our core values which were at odds with slavery. In essence, he made out slavery to be Un-American. This can be best seen in a famous Independence Day speech he gave in 1852 that is often given the title "What to the Slave is the 4th of July?" I was struck as I read his speech how he went out of his way to celebrate the founding fathers, the Declaration of Independence, and the Revolutionary War. For example, he said, "The signers of the Declaration of

The Procedures

Independence were brave men. They were great men too – great enough to give fame to a great age. It does not often happen to a nation to raise, at one time, such a number of truly great men." Douglas also said, "They were statesmen, patriots and heroes, and for the good they did, and the principles they contended for, I will unite with you to honor their memory." And near the end of the speech he drove the point home saying, "I, therefore, leave off where I began, with hope. While drawing encouragement from the Declaration of Independence, the great principles it contains, and the genius of American Institutions..."[5]

But Douglass didn't just praise the core values that birthed our nation. He ruthlessly attacked slavery as a violation of our core values saying:

> What, to the American slave, is your 4th of July?
> I answer: a day that reveals to him, more than
> all other days in the year, the gross injustice
> and cruelty to which he is the constant victim.
> To him, your celebration is a sham; your
> boasted liberty, an unholy license; your
> national greatness, swelling vanity; your
> sounds of rejoicing are empty and heartless;
> your denunciations of tyrants, brass fronted
> impudence; your shouts of liberty and equality,
> hollow mockery; your prayers and hymns, your

Dare To Dream

sermons and thanksgivings, with all your
religious parade, and solemnity, are, to him,
mere bombast, fraud, deception, impiety, and
hypocrisy — a thin veil to cover up crimes
which would disgrace a nation of savages.
There is not a nation on the earth guilty of
practices, more shocking and bloody, than are
the people of these United States, at this very
hour.[6]

What Douglass did, calling Americans back to our core values, has become an inspiration and strategy for Americans fighting violent injustice to this very day. Douglass may not have been the first human rights activist to use this strategy, but he used it so effectively that we remember him today. One of the reasons it was so powerful is because America is a nation whose identity is rooted in our core values from our founding. Our identity is not tied to ethnicity. This isn't the case for many countries around the world whose identities are in some way tied to ethnicity. Russia is of the Russians. Japan is of the Japanese. Ireland is of the Irish. But America is different. America is a nation of liberty lovers. America is about being a place where people can get a fresh start and build their own future. America is about self-determination and individual rights. But most importantly, America is about giving everyone a

The Procedures

fair shake, equality and justice for all! America isn't special because we got it right. Clearly we got it wrong over and over again. And we continue to get it wrong with the widespread and bloody violence of abortion. But America is special because we aspire to get it right from our very founding. So many times we have gone back to our core values to end oppression. We must keep going back to our core values if we are to continue to be Americans, liberty loving people.

111 years after Frederick Douglas gave his "What to the Slave is the 4th of July?" speech, another African-American man, Dr. Rev. Martin Luther King, stood on the steps of the Lincoln Memorial and gave another speech that shook the nation. King's "I Have a Dream" speech also appealed to America's core values in order to end segregation and the violence of the Jim Crow south. His inspiration came from a phrase that was popularized just a few decades earlier, the "American Dream." For so many Americans, the American Dream became a phrase that summed up all of our core values. King understood that and he wanted all Americans to understand that African-Americans were not being included in their dream.

King used the analogy of a bad check to describe America's core values which were being withheld from African-Americans. He preached:

In a sense we've come to our nation's capital to cash a check. When the architects of our Republic wrote the magnificent words of the Constitution and the Declaration of Independence, they were signing a promissory note to which every American was to fall heir. This note was a promise that all men – yes, black men as well as white men – would be guaranteed the unalienable rights of life liberty and the pursuit of happiness. It is obvious today that America has defaulted on this promissory note insofar as her citizens of color are concerned. Instead of honoring this sacred obligation, America has given the Negro people a bad check, a check which has come back marked "insufficient funds."[7]

In King's powerful analogue, America was founded on a promise to defend the natural rights of all its citizens. This promise was represented by a check that was bounced. King had the courage to accuse America of not keeping its word, of bouncing a check. There are very few things that are as valuable to a man as his word. To attack a man's word is to attack his honor and respect. Essentially, King attacked America's honor. He embarrassed us where we so greatly deserved to be embarrassed. And it worked. Real men keep their

The Procedures

promises. And a genuine America keeps its promise to defend the natural rights that our country was nobly founded upon.

King went on to give his people desperately needed hope in the now infamous "I have a dream" portion of his speech. He started with "It is a dream rooted deeply in the American dream. I have a dream that one day this nation will rise up, live out the true meaning of its creed: We hold these truths to be self-evident, that all men are created equal." King went on to describe five pictures of his dream. He was painting a picture of what America could be. He was painting a picture of a liberty worth pursuing even when things seemed hopeless. The picture he painted was powerful, but in its time seemed far-fetched. Could the children of slaves and slave owners really eat a meal together? Would Mississippi really be transformed into an oasis of freedom? Would our nation really judge King's children by the content of their character? King ends his dream with a picture of Alabama, the very epicenter of police brutality and bombings, being a place where black boys and girls and white boys and girls hold hands as brothers and sisters in a mutual human family.[8] Could this really be possible? Or was King just being a silly dreamer?

But what I discovered was that not only was King not a silly dreamer, in fact he didn't dream enough!

Dare To Dream

King described black and white children as part of a human family. But even King didn't dare to dream of white children and black children actually being brothers and sisters. And yet my son is black and my girls are white. And our family isn't even unusual. I know many families just like mine where black and white kids are truly siblings. In his time, King's dream may have seemed far-fetched. If only he could have lived to see my family! While we aren't finished yet in the cause of racial justice, we have transformed further in 50 years than even the Reverend King could have dared to dream.

And yet with all the reasons to celebrate the cause of equality and justice for all, there is even more reason to grieve and be embarrassed. We didn't simply end segregation. Soon after the civil rights movement, we embraced a new violent oppression: abortion. How can we boast equality, after replacing the sweltering heat of segregation with the chilling cold steel of the curette blade? How can we claim inalienable rights, after trading the nooses used to hang black bodies with the vacuum used to dismember tiny fetal bodies? How can we claim to be the people of liberty, after trading the biting teeth of Bull Connor's attack dogs with the sharp steel teeth of the abortionist's forceps? How can we hold up the banner of "life, liberty, and the pursuit of happiness" after replacing the bombs and the burnt bodies of four little girls at the 16th Street Baptist Church with the countless little fetal bodies fatally burned by poisonous saline injection? How dare we utter the word "freedom" while unfathomable numbers of severed arms, legs, heads, and torsos of prenatal children have been incinerated, ground up, and flushed in the name of liberty! We are certainly the most pitiful of nations. We dare to call ourselves "the land of the free" and we enjoy the most lavish prosperity on earth, but then we engage in such bloody and grotesque violence. No nation on earth has less of an excuse than us. We of all people know better!

Despite our hypocrisy, we still believe in inalienable human rights, including life, liberty, and the pursuit of happiness. We don't believe in born rights as if birth gives you rights. We don't believe in viability rights as if some subjective label of viability gives you rights. We believe in human rights which you possess simply by being human. I am daring to dream and am inspired by the dreams of Frederick Douglass and Martin Luther King. I dream of a day when abortion isn't only illegal, but when no one would even consider it an option. I dream of a day when no woman has to choose between her child and her job or school. I dream of a day when we make it easier, not harder, to be a single parent. I dare to dream of the day when every single human, regardless of circumstance, is afforded equal and

Dare To Dream

inalienable human rights and is treated with dignity and respect. And of course I dream of a day when there is a universal consensus that abortion is a violent and unacceptable act. Am I a silly dreamer? Maybe I am, but I dream anyway, just as the Rev. King taught us. I invite you and even dare you to dream with me. If you believe in our core values, this dream is for you. This is our American dream.

The Procedures

Chapter Notes:

1. Julian, Kate. "Why Are Young People Having So Little Sex?" *The Atlantic.* (2018)

   https://www.theatlantic.com/magazine/archive/2018/12/the-sex-recession/573949/

2. Jones, Tyler. "On Jekyll Island, black history remains prominent" *The Brunswick News.* (2017)

   https://thebrunswicknews.com/life/on-jekyll-island-black-history-remains-prominent/article_3272b285-ee8e-5f5a-a7c7-a0683a9bbb6f.html

3. Hergesheimer, E. "Map Showing the Distribution of the Slave Population of the Southern States of the United States Compiled from the Census of 1860"

   https://upload.wikimedia.org/wikipedia/commons/5/5e/SlavePopulationUS1860.jpg

4. Mason, George. "The Virginia Declaration of Rights" *Virginia Constitutional Convention.* (1776)

   https://www.archives.gov/founding-docs/virginia-declaration-of-rights

5. Douglass, Frederick. "Oration, Delivered in Corinthian Hall, Rochester, By Frederick Douglass, JULY 5TH, 1852."

   https://rbscp.lib.rochester.edu/2945

6. Ibid

7. King, Martin Luther. "I Have A Dream..." Speech by the Rev. Martin Luther King at the "March on Washington" (1963)

   https://www.archives.gov/files/press/exhibits/dream-speech.pdf

8. Ibid

NOTE FROM THE AUTHOR

I want to thank you for taking the time to read *The Procedures*. This is a difficult subject to write and read about. It is often unpleasant and even disturbing. I appreciate that you took the time to read about this subject and that you choose *The Procedures*. Regardless of your position on abortion, I hope you've come away from this book better informed and better prepared to make decisions in your own life. If you think that *The Procedures* is a worthwhile read, would you do a favor for me? Would you recommend *The Procedures* to your friends and family? I would also greatly appreciate a positive review on Amazon. We each can do our part in making a less violent society.

Sincerely,

Jordan Warfel

Made in the USA
Coppell, TX
14 January 2024

27694677R00069